KT-599-359

POCKET EINSTEIN

10 Short Lessons in
Renewable Energy

Also in the Pocket Einstein series

10 Short Lessons in Artificial Intelligence & Robotics

10 Short Lessons in Space Travel

10 Short Lessons in Time Travel

POCKET EINSTEIN

10 Short Lessons in Renewable Energy

Stephen Peake

Michael O'Mara Books Limited

First published in Great Britain in 2021
by Michael O'Mara Books Limited
9 Lion Yard
Tremadoc Road
London SW4 7NQ

Copyright © Michael O'Mara Books 2021

All rights reserved. You may not copy, store, distribute, transmit,
reproduce or otherwise make available this publication (or any part
of it) in any form, or by any means (electronic, digital, optical,
mechanical, photocopying, recording or otherwise), without the
prior written permission of the publisher. Any person who does
any unauthorized act in relation to this publication may be liable to
criminal prosecution and civil claims for damages.

A CIP catalogue record for this book is available from the British Library.

Papers used by Michael O'Mara Books Limited are natural, recyclable
products made from wood grown in sustainable forests. The
manufacturing processes conform to the environmental regulations of
the country of origin.

ISBN: 978-1-78929-288-6 in hardback print format
ISBN: 978-1-78929-289-3 in ebook format

1 2 3 4 5 6 7 8 9 10

www.mombooks.com

Designed and typeset by Ed Pickford
Illustrations by David Woodroffe
Front cover illustration by Siaron Hughes

Printed and bound by CPI Group (UK) Ltd, Croydon, CR0 4YY

CONTENTS

INTRODUCTION

Renewable energy is a no-brainer. Ever since I fell in love with physics at school, I've been fascinated by the idea of a world powered by 100 per cent renewable energy: quietly, safely, cleanly, peacefully and equitably. It is a reality for an increasing number of companies, cities and nations. Renewable electricity is now cheaper than fossil electricity, and climate change is making it a future necessity for the whole planet.

Renewable energy powered our lives long before fossil fuels and will do so long after. Many of the scientific and technological breakthroughs underpinning the modern forms of renewables, such as biomass, water, wind and solar power, are more than a hundred years old. But we spent the twentieth century mostly investing in fossilized sunshine instead of the real thing: oil from the early 1900s, nuclear electricity to replace some coal in the 1950s, and natural gas in the 1960s. It was in the decade after the 1973–74 oil crisis that a re-imagining of renewable energy systems occurred as oil prices quadrupled overnight. A long, slow technological fuse was quietly lit.

I was in my teens then, and unaware of an older generation that were already busy explaining the physics of, and digging the foundations for, the explosion of renewable energy that we're now experiencing in the twenty-first century. During the first year of my physics degree at the University of Sussex in south-east England, reactor number four at the Chernobyl nuclear power plant in Ukraine went into meltdown, sending a plume of radioactive material into the atmosphere. A week later, an easterly air flow, helped by some very wet weather, deposited a layer of radioactive caesium-137, iodine-131 and strontium-90 over much of the north of England, including my home town of Bolton and my beloved Lake District. It brought some serious questions to the fore. Did we really need nuclear power? Were there not better, safer ways to power economies? Thanks to the vision of such pioneers as Godfrey Boyle, Dave Elliott, Bob Everett, Michael Grubb, Peter Harper, Amory Lovins, Catherine Mitchell, Walt Patterson, Lee Schipper, Brenda Vale and Robert Vale, modern renewable energy is now a real and compelling choice for our energy future.

The Intergovernmental Panel on Climate Change (IPCC) was established in 1988, just as I was starting a PhD on the topic, which was rather niche back then. Three decades later, and climate change and renewable energy are centre stage in mainstream popular culture around the world. Our need to reach zero carbon emissions by 2050, as set out by the IPCC, means that now, more than ever, we have to be both honest and smart as we mix physics and energy policy together in the climate change pot. We urgently need to stop wasting energy and we need to ramp up our investments in renewable energy by ten to twenty times what they are today if we are to meet this important goal.

Fundamentally, renewable energy is about humans tapping into tiny amounts of planetary-scale sources of solar, geothermal and gravitational energy. This book begins by looking at the scale of energy flows in both the natural world and the human world. Through these short chapters, we'll discover the enormous potential of renewable energy and consider why it is central to limiting dangerous global heating. On our whistle-stop tour of the six major renewable energy resources – solar, wind, biomass, hydro, geothermal and ocean – we'll examine the key technologies, old and new, that are driving them forward and the pivotal role that electricity will play in our energy future.

There is an exciting, renewable, electric, peaceful, prosperous, safer future just up ahead. We just need a little more imagination to get there. Thankfully, imagination is renewable, too.

01 THERE'S NO SUCH THING AS ENERGY

'It is important to realize that in physics today, we have no knowledge of what energy is. We do not have a picture that energy comes in little blobs of a definite amount.'

RICHARD FEYNMAN (1963)

There is no such thing as energy. This might come as a bit of a shock. Einstein's famous equation, $E = mc^2$, tells us that energy equals mass multiplied by the speed of light squared, so there must be such a thing. Energy, it turns out, is much slipperier than our everyday understanding suggests. We might think about paying our energy bills, investment in new energy technologies and the wars that are fought over access to geographical blobs of oil and gas but, intellectually, the scientific idea of energy is mind-bogglingly abstract.

If you were to ask a physicist, 'What is energy?', they would probably explain it as something that exists in various forms and can be converted from one to another. They might

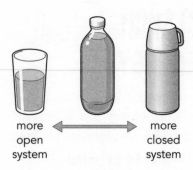

more open system ⟷ more closed system

discuss examples of these conversions, before talking about a very important law of physics: that of the conservation of energy, which is the idea that the amount of energy within a defined boundary (or 'closed system', as physicists would say) remains the same. In other words, energy cannot be created or destroyed.

Energy is a mysterious property that has intrigued and baffled philosophers, mathematicians and scientists for millennia. Variants of the word 'energy' date back to the Greek *enérgeia*, from '*en*' meaning in, and '*ergon*', meaning work. However, none of them were used in the sense of our modern-day scientific understanding of energy as the capacity or potential to do work. The history of exploring the physics of energy is rich in observations about the conversion of energy from one sort to another.

What, in fact, is being conserved is only understood by uncovering the mathematics behind each of the forms of energy in a system. Today, we understand energy at its most fundamental level as mathematics. Before we had uncovered the laws of physics and their various mathematical expressions as they relate to different forms of energy, we had some really fascinating – and sometimes strange – notions of the *thing* that is being conserved.

Dead and living forces

The law of the conservation of energy is held so dear today because it was so hard won. Our journey to the present modern understanding of the idea of energy has taken many interesting twists and turns. It's a 200-year detective story about what exactly is being conserved in a system.

We've been scientifically curious about the idea of heat since ancient times. The idea that things contain heat and that this had something to do with the motion of things too small to see with the naked eye was being discussed by the English philosopher Francis Bacon as early as the sixteenth century. Bacon was one of the first to realize that heat had something to do with the motion of small things, though his observations that 'pepper, mustard and wine are hot' might have suggested he wasn't about to discover an important law of physics just yet. Known as the father of experimental philosophy, his reputedly one and only actual experiment involved stuffing a chicken with snow shortly before he caught a chill and died.

> **Women are the largest untapped renewable energy source in the world.**
>
> NEHA MISRA, co-founder of Solar Sister (2019)

> **❝ Fire made us human, fossil fuels made us modern, but now we need a new fire that makes us safe, secure, healthy and durable. ❞**
>
> AMORY LOVINS (2013)

Hot on his heels, the French philosopher René Descartes understood the world as comprising three types of matter with differing viscosities (liquid thicknesses): fire, air and earth. The Dutch philosopher Christiaan Huygens observed that something he called *calculatrix* was conserved when objects collide. Not a character from *Asterix*, calculatrix is in fact a mathematical quantity related to our modern notion of kinetic energy.

A huge step forward came when the notion of what is being conserved was linked to the concept of force. The German polymath Gottfried Leibniz explored a world of dynamic forces and the distinction between kinetic and potential energy with his theory of *vis viva* (living force) and *vis mortua* (dead force). Isaac Newton gave us our modern notion of mechanics through his famous laws of motion, though this wasn't yet a fully developed 'energy view' of the world. Two early German alchemists, Johann Becher (possibly the first to contemplate the idea of an invisibility cloak) and his colleague, Georg Stahl, understood things to contain a combustible material called phlogiston (from the Greek *phlog*, or *phlox*, for flame). Herman Boerhaave, a Dutch physician known as the founder of clinical medical teaching, later went back to the idea of things containing fire particles and believed that heat was one of four invisible fluids along with electricity, magnetism and elasticity. The Swiss mathematician and physicist Daniel Bernoulli, a member of the prodigious mathematical Bernoulli clan and whose famous equation is associated with the physics of aeroplane wings among other things, then took the thinking a little further, moving us from a world view in which materials and systems harboured mysterious dead and living forces to one of the first truly energy-based understandings of the world. His work helped consolidate the ideas of kinetic, potential and total mechanical energy.

The idea of substances having properties related to heat and to force or work developed side by side. In the nineteenth century, James Prescott Joule, the son of a Salford brewer, was one of the first to publish an accurate calculation of the lock-step relationship between work and heat. In the great age of steam, the ability to do more work with less heat in order to power the Industrial Revolution was as much about capitalism and

> **The quantity of heat capable of increasing the temperature of a pound of water ... by 1 °F requires for its evolution the expenditure of a mechanical force represented by the fall of 772 lb through the space of one foot.**
>
> JAMES PRESCOTT JOULE
> (1850)

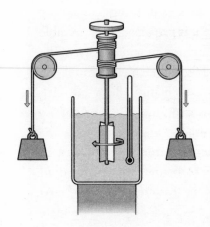

the economy as it was about philosophical insights into nature.

The top of James Joule's gravestone in Sale Cemetery near Manchester is proudly inscribed with the number 772.55. This is his famous 1878 measurement of the mechanical equivalent of heat – the most accurate of its kind – in which he found that this amount of foot-pounds of work (the amount needed to raise a pound (0.45 kg) in weight through one foot (0.3 m) against the force of gravity vertically) must be expended at sea level to raise the temperature of one pound of water from 60 to 61 °F (15.5 to 16.1 °C) in what looks like an early wooden ice-cream maker. This famous result from his paddle wheel experiments, published by the Royal Society in 1850, overturned conventional scientific wisdom that thermal effects were due to the action of a subtle fluid called caloric and established the universality of the law of conservation of energy known as the first law of thermodynamics.

Forms of energy

The first law of thermodynamics is also known as the law of the conservation of energy. It states that in any bounded

DESCARTES AND THE HUNT FOR 'ENERGY'

The concept of energy remains elusive to physicists and philosophers today. There's no agreement when, where or who discovered energy per se, but a good place to start is with the work of the philosopher René Descartes. Most of us might be familiar with his passion for theorizing about mental worlds (he is famous for his 'Cogito ergo sum': I think therefore I am), but he is perhaps less well known for his interest in theories about the physical nature of the world. Descartes was the chief exponent of a new 'mechanical philosophy' of nature that saw gravity, magnetism, colour and smell as something to do with the inner motion and collisions of things – not widely believed at the time due to an unwavering belief in God and his immutable creations. The young philosopher's decision to devote his life to finding a complete system of knowledge came to him on a chilly November night, locked in a room with a new hi-tech Kachelofen – and not, as the myth has it, locked inside the masonry stove itself.

(closed) system, no matter what happens, energy can neither be created nor destroyed. In other words, something we call energy is conserved. This is a particularly important and useful law, especially when attempting to track what happens when energy is converted from one form to another through various complicated physical or chemical processes. No matter the process, we can always get our calculators out and go on an energy hunt, and somewhere, in some form or other, we'll find the same amount of energy as we started with.

This is by no means intuitive, especially from the point of view of human observations of the array of dynamical interactions that go on in the world around us, where energy seems to slip away eventually to nothing (we will come to that shortly). But this does help explain why, despite its name, the first law was in fact only discovered after the second (see page 16)!

One of the ways in which the spirit of seventeenth-century natural philosophy lives on, rather nostalgically, is in the still modern and ubiquitous idea that energy comes in different 'forms'. Modern curriculums at all levels of education and all over the world hunt and collect the different forms of energy, just as our early chemists and physicists tried to make sense of the sometimes strange and mysterious properties inside

> **"Energy" was not simply waiting to be discovered, like a palaeontologist might find the first ichthyosaur or a prospector stumble across the Koh-i-Noor diamond.**
>
> JENNIFER COOPERSMITH
> (2010)

the materials they observed – the calculatrix, the phlogiston and so on.

Intriguingly, our modern energy theory continues to divide the world into the two basic elementary forms: kinetic energy and potential energy. The insights of Gottfried Leibniz in the seventeenth century and other early physicists on living and dead forces remain astonishingly prescient today.

Motion is kinetic energy. Position relative to a force is potential energy. Real-world systems are always a mixture of the two. However, for the purposes of explication, we like to choose specific examples to illustrate the different forms. Most of us will have been told at school that stored energy is always potential energy, such as the water in a reservoir – but this isn't quite true. The kinetic energy of a large ocean wave is a large (temporary) store of energy, too.

Types of kinetic energy include: the mechanical energy of large objects (wind, water and tidal current turbines, planes, trains and automobiles); the thermal energy of microscopic particles in fire and fluids (the jiggling of molecules and atoms); the flow of electrical energy (electrons) in electric circuits; the flow of sound through air; and the energy of electromagnetic radiation from short wavelengths around 500 nanometres (for example, visible light, which is roughly the length your fingernails grow in eight minutes at 1 nanometre per second) to long wavelengths (for example, radio waves of a 100 metres, 328 ft or so).

Types of potential energy include gravitational energy of water held in a hydro scheme; electric potential energy in batteries; magnetic potential energy in electrical motors; chemical energy in food; elastic energy in springs; nuclear energy in uranium (or indeed, in principle, any mass through Einstein's energy–mass equivalence – see Chapter 2 for more on this).

> Because we use a hundred and ten times as much coal as our ancestors, we believe ourselves a hundred and ten times better, intellectually, morally and spiritually.
>
> ALDOUS HUXLEY (1928)

The main energy conversions associated with renewable energy technologies include: solar (radiant energy to electrical energy), wind (mechanical energy to electrical energy), hydro (mechanical energy to electrical energy), biomass boilers (chemical energy to thermal energy), tidal barrage (gravitational energy to potential energy to mechanical energy to electrical energy), wave (mechanical energy to electrical energy) and geothermal (thermal energy to electrical energy).

The standard scientific unit for measuring energy and work is dedicated to James Joule. It's a relatively tiny amount of energy at the human scale. Our daily food energy requirement, for example, is about 8 million joules. A joule is roughly the amount of energy required to raise an apple (100 grams) through 1 metre (3.3 ft) of height.

Eat the apple and you will consume about 200,000 joules. Power is the rate at which energy flows or work is done, and a rate of one joule per second is a watt, named after the steam engineer James Watt. Humans consume energy at around 100 joules per second, or 100 watts.

When thinking about the state of our global economies and the transition away from fossil fuels towards large-scale renewable energy systems, the amounts of energy are relatively large. As we'll see in Chapter 2, the sun is the source of almost all our renewable energy and sets the scale of measurement. Typical measures of energy and power for different types of renewable energy involve thousands (10^3 kilo), millions (10^6 mega), billions (10^9 giga), trillions (10^{12} tera), quadrillions (10^{15} peta) and quintillions (10^{18} exa) quantities of joules (when measuring energy stores) or watts (when measuring energy flows).

We are used to the idea of kilo (as in kilograms) and the term mega is commonly used to mean 'big' (as in megabucks), but the scientific notation for giga, tera, peta and exa are not intuitive. To remember the orders of magnitude correctly (mega-giga-tera-peta-exa), it might help to keep this little mnemonic in mind: **M**illions of **G**reen **T**urbines **P**ower **E**lectrons.

Entropy and exergy

The neat definitiveness of the first law of thermodynamics is rather comforting, but isn't always that useful. It doesn't tell us what happens to the energy in a system as physical and chemical processes take place over time. The second law is much more complicated and much less easily grasped than the first; it tells us that in every real-world energy transformation, something is used up. But what is used up and how is this measured? It is certainly not energy: the first law says it can't be.

The second law states that energy transformation processes are not random but have a tendency to maximize something called entropy. Entropy is a measure of the disorder in a system. A more highly ordered system (books arranged neatly and in alphabetical order on a desk) has a lower entropy than a more disordered system (books all over the desk in no order).

Where there is a difference in temperature between objects or systems that are connected to each other, things will always come into thermal equilibrium – they like to settle to their lowest common denominator. For example, ice in a drink melts because the ice warms and the drink cools until there is thermal equilibrium between the two. The second law tells us that heat flows from the hotter to the cooler – and never the other way around. Our trade winds and ocean currents are flows of air and water redistributing the sun's heat energy from where it concentrates around the equator towards the poles.

It is not just thermal systems that don't like differences. In general, the world is driven by unstoppable forces sliding it to sameness – to homogeneity. The world is full of examples of an overall tendency for nature to slide towards disorder when left to its own devices. In the absence of purposeful life-forms (plants, animals, people), things tend to slide into uniform chaos. Highly engineered, manufactured and ordered systems such as libraries, skyscrapers and cities don't happen by accident – they have low states of entropy that only increase the moment you leave them alone. Gardens are also a good example: leave them alone and they tend to disorder. A barrel of oil contains a rich mixture of long-chain hydrocarbons packed with potential chemical energy to be sorted and converted into all sorts of different chemicals used to produce energy and/or plastics or chemicals. The heat leaving the refinery or delivery tanker or car engine and exhaust system is a much higher entropy and a less useful state.

The flipside of this idea called entropy is a concept known as exergy. In real-world energy transformation processes, entropy increases and exergy is used up. Exergy is the amount of energy that is available to be used to

> **❝ Nothing in life is certain except death, taxes and the second law of thermodynamics. All three are processes in which useful or accessible forms of some quantity, such as energy or money, are transformed into useless, inaccessible forms of the same quantity. ❞**
>
> SETH LLOYD, Professor of Mechanical Engineering and Physics, MIT (2004)

perform useful work. This is a tricky concept to get your head around. One gigajoule (1 billion joules) of energy is 278 kWh of electricity (three weeks' consumption for an average UK household) or 36 kg (79 lbs) of coal, or the extra energy stored when we raise the temperature of a small 24-tonne garden swimming pool by 10 °C (18 °F). They all represent 1 GJ of energy, but it is not all available to do useful work.

We can't get 100 per cent of the energy in coal, or that stored in a swimming pool, to do meaningful work. Sometimes we refer to the quality of energy as 'high' or 'low' grade. Heat, for example, is the microscopic random motion of particles, whereas work is a more macroscopic ordered motion of particles.

Heat is therefore less orderly and lower grade. Work is more orderly and higher grade. Electricity (a form of stored work) is a higher-grade form of energy – it has potential to do all kinds of work and can be transformed completely into thermal energy. We can heat the pool electrically.

However, thermal energy is lower grade and cannot be transformed 100 per cent back into electrical energy (or any other form). Even with the best modern combined cycle gasification turbine power station, we need about 2 GJ of gas to make 1 GJ of electricity.

The second law tells us we can't play around indefinitely converting energy from one form back to another and back again without losing something along the way. It is why the hypothetical idea of perpetual motion machines that can carry on working indefinitely without an energy source has yet to be demonstrated, despite the many valiant and imaginative efforts of inventors.

There are different ways of expressing how the quality of energy is changed during energy transformations. With every energy transformation there is:

- An increase in entropy: a loss of order/ heterogeneity/complexity and an increase in disorder/homogeneity/sameness.
- A decrease in exergy: a loss of usefulness/a degradation of the quality of the energy in the system.

To generate energy, we have to reorder the world. Or put another way, whenever the world reorders itself, energy is generated. The reordering can be at the nuclear and atomic level (nuclear fission/fusion that powers the sun), or at the molecular level (physical friction and chemical reactions such as complex hydrocarbons combining with oxygen to create energy, water and carbon dioxide), or at macroscopic levels (changes in the pressure, volume or temperature of solids, liquids and gases that power engines).

Our understanding of energy comes from some very imaginative origins, and while we have settled a great deal of the mathematics, the fact is that energy is not a simple 'thing'. Yet it is a critical part of the future of our world. With a burgeoning human population attempting to wean itself off three centuries of addiction to fossil fuels, yet hungry still for energy (electricity in particular), there remain some quite strange economic, technical and social ideas about what energy is, why it is significant and how we might go about managing it.

02 NOTHING IS TRULY RENEWABLE

'The sun provides five thousand times more energy to the earth's surface than our total human demand for energy.'

MARTIN REES (2018)

There's nothing remotely renewable about the Big Bang. Well, probably. Where did the energy come from in the earliest moments of the universe's hot, dense phase 13.8 billion years ago? The short answer is, we don't know. Did it come from nowhere? Did it somehow already exist? As we continue to postulate how energy affects the universe at the largest cosmological scales, there is still a lot of physics to be discovered.

Perhaps it is something to do with dark energy. We have only fairly recently calculated how much of it there is from the physics of the expansion of the universe. About 68 per cent of the universe is dark energy, 27 per cent dark matter and 5 per cent is what you and I would call ordinary matter: stuff such as the earth, the planets, the sun and the stars and the material you are reading this book on.

The notion of a rapid expansion of the universe after the Big Bang is a low-entropy to high-entropy story. Or put another way, the Big Bang is the ultimate example of thermodynamics' 'arrow of time' at work.

Scientists believe that the early universe was an even distribution of hotness – a low-entropy state – from which it has been gradually expanding and cooling. There are several theories about how the cosmos might end. Most suggest there are a few billion years left. The Heat Death theory, in which the universe continues to expand for ever, eventually creates a state of thermal equilibrium and in doing so maximizes entropy (low exergy), so nothing interesting ever happens again. Then there's the Big Crunch theory, where the universe stops expanding and collapses into a black hole. There is also the theory of Eternal Inflation, which suggests we have around another 5 billion years left until *time* itself ends, in this part of a group of universes known as the multiverse.

But long before any of this might happen, the future of our own solar system and all life on this beautiful earth

> ❝ In the beginning the universe was created. This has made a lot of people very angry and been widely regarded as a bad move. ❞
>
> DOUGLAS ADAMS,
> *Restaurant at the End of the Universe* (1980)

will be intimately connected with the fortunes of one star in particular, our beloved sun.

Star of the show

Stellar radiation drives the vast majority of the earth's energy transformations and, most importantly, is critical to photosynthesis, the basis of life on Earth. The sun is a gigantic thermonuclear power plant in the form of a giant ball of plasma over a hundred times wider than our planet, composed mainly of hydrogen and helium. Every second, it converts about 4 million tonnes (4.4 million US tons) of matter into energy by fusing around 600 million tonnes (661 million US tons) of hydrogen into 596 million tonnes (657 million US tons) of helium. The energy escapes in all directions in the form of a flux of electromagnetic energy.

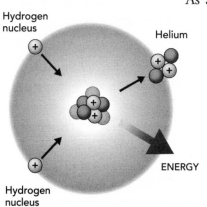

Hydrogen nucleus

Helium

ENERGY

Hydrogen nucleus

As a relatively unremarkable 'G2 dwarf' star, the sun is just over halfway through the main part of its thermonuclear life, steadily turning the hydrogen into helium. Like any other kind of battery, it has a finite lifetime and will eventually become exhausted. It is about 40 per cent brighter than it was just after it was formed 4.6 billion years ago and will eventually become a thousand times brighter still, gobbling up the earth as it balloons into a red giant before collapsing into a white dwarf and then finally into a cold inert star 'skeleton' we call a black dwarf.

The good news is that for the next 1 billion years or so, before the sun becomes uncomfortably bright, it will continue to drive and support life on Earth in all its myriad forms, including the ambitions and prospects of our species, *Homo sapiens*, with pretty much all the energy needs we care to dream up. Well, in theory, at least. As we move away from fossil fuels to a renewable energy future, those needs will be met increasingly directly from the sun. A deeper look using a little maths and physics quickly reveals the enormous flows and stocks of solar energy stored in the land, oceans, ice and biosphere. Let's pop some energy goggles on and take a look.

The Earth–atmosphere energy balance

We live in a world awash with energy. We don't need much, if any, physics know-how to appreciate the scale of its flow through the natural world. Our day-to-day weather systems and regional climate differences are all part of the massive redistribution of the sun's heat energy from the warm equator towards the cool polar regions. The trade winds, sea currents and jet streams are the planet obeying the second law of thermodynamics in the search for thermal equilibrium.

In the last fifty years or so, we have come to know the earth as a complex interplay of different sub-systems. earth systems science, as we call it, is about the exchange of energy and material between five 'spheres': the atmosphere (the thin layer of gases above the surface), biosphere (all living things), hydrosphere (fresh and salt waters), cryosphere (ice and glaciers) and lithosphere (rocks, volcanoes). Thanks to its atmosphere, the earth manages to maintain a relatively convivial global average surface temperature of about 15 °C (59 °F). Without its atmosphere, the earth would have an average surface temperature of -18 °C (-0.4 °F), 33 °C (59 °F) cooler.

The earth is an open energy system: it receives energy from the sun and reflects roughly the same amount back into space. The flux of solar electromagnetic energy radiates outwards in all directions from the sun. As a relatively small object, many miles from the sun, the earth intercepts just a tiny disc-shaped fraction of this – in fact, the sun emits around 2.2 billion times the share of energy that the earth catches.

> **We are star stuff harvesting sunlight.**
>
> CARL SAGAN (1980)

At any moment, the average power of the sun's energy falling on the earth is 1.74×10^{17} W, or 174 PW. On average, the flow of solar radiation energy passing through a square metre at the top of the atmosphere is roughly 1370 W/m^2 (+/- 3.5 per cent due to the earth's orbit of the sun). Although this figure varies a little over time, we call it the solar constant. The surface area of the earth is four times the area of the disc that the sun 'sees'. Dividing by four, we get an average power density (power per unit of area) of incoming solar radiation at the top of the atmosphere of around 342 W/m^2. If the earth doesn't reflect roughly this amount of energy back into space, over time it would heat up or cool down. In fact, rising greenhouse gas concentrations from burning fossil fuels, deforestation and farming has tipped the earth into a phase of global heating, as we will see in Chapter 3.

The electromagnetic energy is delivered in a spectrum of wavelengths from short gamma waves to long radio waves. About 9 per cent of this energy is ultraviolet, 39 per cent is visible spectrum and 52 per cent is infrared. The energy arriving from the sun contains relatively more short-wave radiation energy than the earth reflects back, which contains relatively longer wave length (infrared) radiation.

Around 30 per cent of incoming solar radiation is immediately reflected back to space by clouds and the

THE 'WORLDWATT'

In 2019, our global economy consumed all forms of primary energy at a rate of 19,000 billion joules per second or about 0.019 PW, or 19 TW, or 19,000 GW. To make it easier to compare flows of renewable energy in nature with the current rate of global primary energy consumption, we can simplify these sometimes-confusing energy units. We can call the rate at which the world uses all forms of primary energy, 1 'worldwatt'. This quantity is a very handy yardstick as we take a tour around the earth's systems and their renewable energy flows in the chapters to come. The 87 PW of radiation received from the sun (which is simply the 170 W/m^2 over the whole surface) is about 4,500 worldwatts. The availability of radiant solar energy is many thousands of times greater than our wasteful fossil-fuelled global energy economy. Put another way, in just under two hours (115 minutes) the sun delivers the same amount of primary energy that the global economy uses over the course of a year.

earth's surface (which is why it looks so lovely from space), and another 20 per cent bounces around the atmosphere making itself useful in the greenhouse effect that keeps our planet warmer than the moon. By the time the rest has made its way through our thin atmosphere it delivers

ANCIENT FOSSILIZED SUNLIGHT

The coal that powered the Industrial Revolution was formed around 300 million years ago in a geological period of time known as the Carboniferous. Oil and gas were deposited in the swamps of the later Jurassic and Cretaceous ages (about 65–200 million years ago). During these relatively cool and wet periods of the earth's climate history, trees and marine organisms were able to sequester vast amounts of carbon from the atmosphere to the land and oceans, resulting in the oil, gas and coal deposits we use today. Fossil fuels are carbon captured by plants and marine animals and stored by geology; they are ancient sunlight. While the easy-to-get-at coal, oil and gas has nearly all gone, there is still a lot of carbon stored in the ground. Our best guess is that, so far, we have burned through approximately 10 per cent of what's actually there. A rough estimate for the remaining fossil-fuel energy reserves is 200 zettajoules (a zettajoule is one thousand exajoules or 1,021 J). To put things in solar perspective, all the carbon left stored on Earth is about twenty-six days of the total solar radiation that reaches the earth's surface.

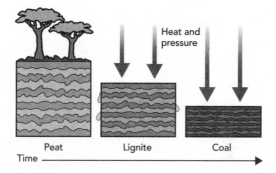

a global yearly average power of 170 W/m² (about half the average at the top of the atmosphere). This adds up to 87 PW of energy delivered to the earth's surface each year. At sea level, on a clear day with the sun directly overhead, solar radiation will instantaneously reach above 1,000 W/m². The interaction of land masses, seas, mountains and clouds can create massive variations in solar flux reaching the surface.

The earth's axial tilt, season variations, clouds, oceans and continents all affect the amount of solar radiation reaching the earth's surface. Cloudiness can cause amazing local differences in annual averages: a classic example is the difference between average solar radiation of the cloud-shrouded Oahu (150 W/m²) and Pearl Harbor (250 W/m²), just 15 km (9.3 miles) apart.

The solar radiation that reaches the earth is at the heart of most physical and chemical processes that drive life on Earth, with the addition of a little internal heat and a bit of gravity. We live off the sun. Until relatively recently, our ancestors lived 100 per cent renewable-energy lifestyles based on the power of the sun, eating plants, using animal power, as well as a bit of water and wind power. In contrast, the complex high-energy society of today is reliant on enormous material flows of fossil fuels, a great deal of which we waste. In terms of the earth's overall energy balance, as well as the direct solar flux, there is a smaller contribution from gravity, the earth's latent heat and its own thermonuclear energy releases.

In theory, there are vast amounts of all kinds of renewable energies available to us, but in practice the amounts are considerably less when we take into account the economic costs of capturing these energy flows as well as competition over the use of land for other purposes – for example, growing food. We also need to be wary of pure physics perspectives on measuring stocks and flows of energy in nature. Energy is everywhere, and we could calculate the energy contained in molecules, in a large cumulus cloud, in a volcanic eruption, in the ocean or even perhaps the kinetic energy of the rotating Earth. But catching it and putting it to use is another matter.

Renewable energy potential

In practice, we will only ever be able to capture a small fraction of the different flows of renewable energy that we find in nature. Technical, economic and political feasibility assessments of the size of that fraction are complex and highly uncertain. A Special Report, *Renewable Energy Sources and Climate Change Mitigation*, published by the Intergovernmental Panel on Climate Change in 2011, reviewed the academic and technical studies to make some rough estimates. The results are startling. Taking a conservative mid-estimate of the numbers in the literature, we see that the global technical potential of different renewable sources adds up to 46 worldwatts. There is a definite and reasonable prospect of humans harnessing 1 worldwatt from 100 per cent renewable energy in the future. To avoid dangerous climate change, the sooner the better.

Ordered from largest to smallest, here's a snapshot of the renewable energy potential of different resources (in worldwatts). In 2018, we used 0.13 worldwatts of renewables.

TECHNICALLY CAPTURABLE GLOBAL RENEWABLE ENERGY RESOURCES (WORLDWATTS)

	Low estimate	High estimate	Mid-estimate
Solar	2.6	83.2	42.9
Geothermal	0.2	2.4	1.3
Wind	0.1	1.0	0.6
Biomass	0.1	0.8	0.5
Ocean	0.0	0.6	0.3
Hydropower	0.1	0.1	0.1
Total	**3.2**	**88.0**	**45.6**

Solar energy

By any measure, the global solar resource available for humans to tap is a massive 43 worldwatts – many times our current (and likely future) energy needs. The closer a country is to the equator, the higher its solar potential. There are four major types of solar energy: (1) solar thermal; (2) solar photovoltaic electricity; (3) concentrating solar power; and (4) solar fuels. Choices about how to use our land (or oceans) for the efficient and economic extraction of solar energy will involve making choices between some of these competing technologies.

Geothermal

The earth's heat flux is about 40 TW (2 worldwatts) and about half of that (or roughly just over 1 worldwatt) may be technically available to tap. Most, but not all, of the energy that shapes the planet comes via the sun. A smaller contribution comes from the fact that the earth is itself a giant heat battery. The heat energy it stores is about 100 billion years' worth of our current global annual energy consumption. Ever since its formation, the earth's core has been slowly cooling. The leaking of this primordial heat over billions of years has driven the movement of the earth's plates, powering sea-floor spreading, mountain-building and generally moving continents around the place. Another source of the geothermal heat flux is the radioactive decay of uranium, thorium and potassium. We are not quite sure, but we think about half the flux might be primordial heat and half might be related to radioactivity.

Wind

Just 1 per cent of incoming solar energy creates the thermal energy powering the patterns of low and high pressure that drive the world's wind energy systems. These flows are driven by the redistribution of tropical heat in the form of trade winds. The global wind energy resource is around 1 PW (about 50 worldwatts), though only a fraction of this in reality could ever be harnessed. Onshore wind potential is about four times greater than offshore potential, both coming to about 0.5 worldwatts.

Biomass

It is estimated that total global primary biomass production (above ground) is equivalent to about 2–3 worldwatts. Only a small portion of this is available to safely use for energy production because of our urgent need to protect biodiversity by preserving our ecosystems and at the same time supporting other uses of it such as forestry and food production. There are three basic types of biomass energy: (1) direct by-products from food and fibre production in agriculture and forestry; (2) indirect waste streams from food and forestry; and (3) plants specifically grown for energy production. A reasonable guess for biomass energy potential is 0.09 PW, or just under half a worldwatt.

> **Calculating how much carbon is absorbed by which forests and farms is a tricky task, especially when politicians do it.**
>
> DONELLA MEADOWS
> (2000)

Ocean

There are six types of ocean energy: (1) wave energy; (2) tidal range; (3) tidal current; (4) ocean current; (5) ocean thermal energy conversion; and (6) ocean osmotic (salinity gradients). Marine biomass farming is usually thought of as bioenergy, while submarine heat vents are thought of as part of geothermal. Tidal current is the only other source of renewable energy (along with geothermal) that is not directly related to the solar flux. Instead, they are related to the gravitational energy of the earth-moon-sun system. The tides give us a glimpse of the enormous inertial size of the earth. Calculations suggest that there is about 3 TW of tidal friction. This is about a sixth of a worldwatt and its effect is to slow the earth's rotation 1.5 thousandths of a second every 100 years. The total ocean resource potential is around one-third of a worldwatt.

Hydro

About half of all the solar radiation reaching the earth's surface is used to evaporate water from the oceans and land to drive the global hydrological cycle. Hydro (along with wind and wave power) is sometimes called 'indirect solar energy'. There are two types of hydropower: (1) run of the river and (2) reservoir storage. Hydropower is a mature renewable energy technology and could possibly quadruple over the next decades to about 10 per cent of today's primary energy consumption (0.1 worldwatts).

It is clear that we are surrounded by enormous flows of renewable energy. We can see, feel and hear the evidence of it everywhere. Renewable energy has been flowing, bumping, storing and leaking for epochs, all completely naturally. And then along we came with our pickaxes, spades, pipes, whirly machines and steam engines, digging up the past or catching a bit of water, wind or sunlight and putting it to work. For almost all of modern human existence (around 220,000 years) – including the last 10,000 years of civilization, discovery and invention (the Holocene) – human societies have been powered by renewable energy flows: mostly wood and biomass in particular, but also water and wind power. Only in the last 500 years have human societies evolved technologies to tap into fossilized stores of coal, oil and gas. Globally, the easy sources of cheap oil, gas and coal

> **Once the initial investment has been made in renewable energy infrastructure, nature provides the raw materials for free.**
>
> NAOMI KLEIN (2014)

are limited and, in some places, running out. In addition to this, political commitments to reach net zero-carbon emissions to prevent dangerous climate change mean that we are at the end of a period where modern societies have gorged themselves on large flows of concentrated ancient fossilized sunlight to power their energy demands. The age of fossil-fuel combustion is coming to a close and a bright new age of renewable-powered electricity and heat is on its way.

03 THE CLIMATE CLOCK IS TICKING

'No terrestrial civilization can be anything else but a solar society dependent on the sun's radiation.'

VACLAV SMIL, *Energy and Civilization: A History* (2017)

Fossil fuels are on their way out and a new era of renewable energy has begun. Historically, this is par for the course. The development of modern society has been powered by a series of energy transitions. As hunter-gatherers, we – and the animals we ate – lived off plant and animal biomass. Early civilizations learned how to harness animal, water and wind power in the form of the ox, watermills and sailing boats. Coal, as well as charcoal, for higher temperature heat was used as early as the Chinese Han dynasty (about 100 BC) and, in Europe, the earliest use of fossil fuel dates back to 1100.

By the thirteenth century, coal was being shipped between countries in Europe, and 300 years after this, coal mining was widespread in Europe, in part because of fuel-timber shortages due to the ever-increasing demand for iron (which requires charcoal production from wood), and to a voracious and fast-growing ship-building industry. The Industrial Revolution, between 1760 and 1860, saw the decline of human- and animal-powered manual labour in farming, industry and manufacturing. It was a major transition away from these forms of energy, as well as water and wind power, which had facilitated our survival as a species, to the use of fossil fuels.

The Industrial Revolution invented new ways of producing and using fossil fuels, such as coal, oil and gas. It was powered by such inventions as the steam engine and the internal combustion engine, which could convert raw fossil fuels into powerful and concentrated sources of mechanical power and heat. A second great energy transition took place with the production of electricity from hydroelectric sources and steam turbines in the late nineteenth century. We are now in the midst of a third great energy transition as we begin the switch from fossil-fuel-based electricity to modern renewable electricity.

> Public support for combatting climate change, for energy efficiency and for renewables, means politicians are democratically obliged to drive a system transformation.
>
> CATHERINE MITCHELL,
> Professor of Energy Policy,
> University of Exeter (2017)

TECHNOLOGICAL INVENTIONS POWERING ENERGY TRANSITIONS

Modern *Homo sapiens* control fire	300,000 BC
Stone oil lamps	15,000 BC
Tin and copper smelting	6000 BC
Sailing on the Nile	5000 BC
Ox-pulled sledge, Mesopotamia	3000 BC
Egyptian oared boats	2500 BC
Horse-drawn vehicles, Egypt	2000 BC
Passive solar, Greece and China	1000 BC
Coal use in Greece	320 BC
Wind-powered pumps, China, Persia, Middle East	200 BC
House heating with coal, China	100 BC
Waterwheels, Greece and Rome	100 BC
Charcoal for steelmaking	AD 1
Glass windows, Roman Egypt	AD 100
Windmills, Europe	AD 1150
Coal mining in Europe spreads rapidly	AD 1640
Newcomen's steam engine	AD 1690
Savery's 3.75 kW mine engine	AD 1712
Intensive European canal system	AD 1750
Watt's steam condenser	AD 1769
Waterwheel-powered factories	AD 1770
Argand oil lamps	AD 1780
Volta's electric battery	AD 1800
Steam ships	AD 1800

Stephenson's Rocket	AD 1829
Francis' electric water turbine	AD 1847
Oil drilling, Pennsylvania	AD 1859
Otto's four-stroke engine	AD 1876
Swan's filament light-bulb	AD 1878
Edison Electric Light Station	AD 1882
Parsons steam turbine	AD 1884
Benz's first car	AD 1885
Tesla's electric induction motor	AD 1888
Diesel's engine	AD 1892
Wright Brothers' first flights	AD 1903
Korn's PV cell	AD 1905
Gas from coal (Fischer–Tropsch)	AD 1910
Crude oil cracking	AD 1913
Vestas Wind Systems founded, Denmark	AD 1945
Bell Labs' first silicon PV cell	AD 1954
Calder Hall, UK, first nuclear power station	AD 1956
North Sea natural gas first brought ashore	AD 1967

The carbon cycle

As we saw in the last chapter, coal, oil and gas are part of the earth's natural carbon cycle. They are formed from the remnants of dead plants and animals and cooked under pressure over millions of years. They are tangible evidence of the natural exchange of carbon between the different spheres that make up the earth's sub-systems.

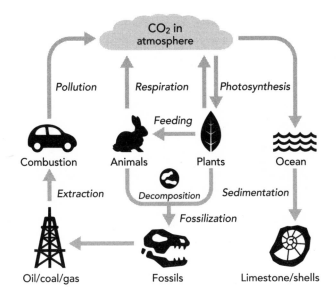

If you go on a carbon hunt around the earth's systems, it's easy to become overwhelmed by the numbers and the sheer scale of quantities of carbon stored in one form or another. While the earth's atmosphere contains just under 0.5 per cent concentration of carbon dioxide, this actually amounts to 800 billion tonnes (882 billion US tons) of carbon (about double the amount of all fossil fuels we have burned in history). There is around 500 billion tonnes (551 billion US tons) in the world's plants, 2,000 billion tonnes (2,205 billion US tons) in the soils and 1,700 billion tonnes (1,874 billion US tons) trapped in permanently frozen land known as permafrost. The first 100 metres (328 ft) of surface ocean contains about the same amount, while the deep ocean stores sixty times this amount. There are about 1,600 billion tonnes (1,764 US tons) in proven coal, oil and gas reserves but there are, in addition, vast quantities of carbon contained in the earth's sedimentary rocks and mantle. So, there's a whole lot of carbon around the earth.

These large carbon stocks are the result of billions of years of Earth history. Generally, the earth's systems are more or less in balance on shorter time scales, these stocks changing very slowly. But our annual emissions of carbon due to burning fossil fuels are considerable. While much carbon is locked away in long-term storage, some of it is whizzing between land plants, soil, animals, the atmosphere and the ocean surface waters pretty quickly. We are pumping out more than the land and oceans are able to take up each year. The result is that we are parking more carbon dioxide in the atmosphere and so its concentration is increasing.

When we dig up fossil fuels, burn them and release carbon dioxide into the atmosphere, we are interfering with the earth's natural carbon cycle. Individual carbon dioxide molecules last just a few short years in the atmosphere before being grabbed by a plant or sucked into the ocean surface, possibly before going around the cycle again. When we add relatively large quantities of carbon dioxide to the atmosphere in a relatively short period (by natural carbon cycle standards), the overall effect on the entire system can

FOSSIL CARBON LEAVES A TRACE

When a plant or animal dies and sinks into the ground, into a swamp or to the bottom of a lake or ocean, the carbon it is made of is locked up. There are three different types of carbon that occur naturally, known as carbon isotopes: ^{12}C, ^{13}C and ^{14}C. The different isotopes have slightly different molecular masses (they contain six, seven and eight subatomic particles called neutrons respectively) and this affects the rates at which they are taken up in different carbon reservoirs over time. About 99 per cent of carbon is ^{12}C, 1 per cent ^{13}C and a tiny fraction is the unstable ^{14}C. We can use isotopic ratios as a sort of carbon fingerprint. Plants prefer the lighter ^{12}C over the heavier ^{13}C. The $^{13}C/^{12}C$ ratio is about 2 per cent lower in plants than it is in the atmosphere. Since fossil fuels are ancient plants, they too have less ^{13}C. When we dump 440 gigatonnes (485 US gigatons) of fossil carbon into the atmosphere we are diluting the normal background concentration of ^{13}C to even less than its standard 1 per cent. Isotope geochemists can analyse tree rings, mud cores and ice cores to look back over the last two or three centuries and have observed a decline in atmospheric ^{13}C. This indicates that the additional carbon in the atmosphere has come from fossil fuels.

be some long-lasting carbon 'traffic jams'. For a 100-billion-tonne (110 billion US tons) pulse of carbon emitted into the atmosphere at today's concentrations, atmospheric carbon dioxide levels will remain roughly 25 billion tonnes (28 billion US tons) higher after a thousand years. After a few decades, the oceans will have taken up 60 billion tonnes (66 billion US tons) and the land about 15 billion tonnes (17 billion US tons).

The concentration of carbon dioxide in the atmosphere has increased by just under 50 per cent since the start of the Industrial Revolution and is now around 412 parts per million (just under half a millilitre's worth), rising by about 2 ppm per year. That rate of increase may not sound a lot, but the rate is about 150 times faster than anything we have seen in the last few million years.

We know that the earth's atmosphere helps make the planet 33 °C (59 °F) warmer than it would otherwise be. It does this because some atmospheric gases intercept the incoming flux of short-wave radiation and re-emit the energy at longer

> **The human influence on climate change is clear and dominant. The atmosphere and oceans are warming, the snow cover is shrinking, the Arctic sea ice is melting, sea level is rising, the oceans are acidifying.**
>
> CORINNE LE QUÉRÉ,
> Professor of Climate Science
> and Policy, University of
> East Anglia (2013)

THE KEELING CURVE

The Keeling Curve is a daily record of atmospheric carbon dioxide concentration measured at the Mauna Loa observatory in Hawaii and maintained by Scripps Institution of Oceanography at UC San Diego. It dates back to 1956 when Charles Keeling began taking measurements and is the longest-running continuous measurement of its kind in the world. It's an iconic wiggly line that increases more or less linearly from an average of 315 ppm in the late 1950s to about 415 ppm in 2020. The annual wiggles show how plants in the northern hemisphere's spring and summer breathe in CO_2 and, in the winter, breathe it out. Atmospheric CO_2 concentrations were pretty stable over the last 10,000 years at around 260–280 ppm, until they started to take off in around 1850. The curve is a direct measurement of the additional 275 billion tonnes (303 billion US tons) of carbon dioxide that we have dumped into the atmosphere as a result of burning fossil fuels.

wave lengths. They cause a temporary heating effect – the greenhouse effect – and we call them greenhouse gases.

Water vapour and carbon dioxide are very powerful greenhouse gases. There are about 13,000 cubic kilometres (3,120 cubic miles) of water dispersed as vapour in the atmosphere, and atmospheric water vapour is, of course, natural. It is an important part of the water cycle and has been around on Earth for over 3 billion years.

The additional carbon dioxide, methane, nitrous oxide and some other industrial gases that we have pumped into the atmosphere since the start of the Industrial Revolution cause additional global heating or an enhanced greenhouse effect, and we call them 'anthropogenic greenhouse gases'. Water vapour, depending on its atmospheric concentration, is responsible for two to four times the greenhouse effect of carbon dioxide. Together, these gases have been busy trapping short-wave solar radiation, becoming energetically excited and then vibrating longer-wave infrared radiation back into the atmosphere.

Greenhouse gases

Warming effect

As we saw in Chapter 1, things like to get to thermal equilibrium (maximum entropy) and, left to its own devices, the earth can cope with a pulse of extra carbon dioxide (or methane etc.). There would be a temporary warming – then systems would come back into balance. The problem is that the industrial pulse hasn't stopped.

It's been growing steadily since the start of the Industrial Revolution and globally shows little sign of slowing down, let alone declining (it's too early to say whether the disruption and innovation caused by the Covid-19 pandemic will result in long-lasting carbon and energy savings). So, as the years roll by, we burn more fossil fuels, the atmospheric concentration of carbon dioxide increases and the atmosphere traps more heat.

Our best guess is that for a doubling of CO_2 from 280 ppm (the pre-industrial concentration) to 560 ppm, the global average surface temperature would increase by 3 °C (5.4 °F) (with a 95 per cent chance of it being somewhere in the range of 1.5–4.5 °C (2.7–8.1 °F). Scientists call this estimate the climate sensitivity. Not all scientists agree what this figure should be – some climate models run colder (lower climate sensitivities) while others run hotter (higher climate sensitivities).

About 90 per cent of the extra heat that the atmosphere has trapped is transferred to the oceans – the ocean surface is very good at absorbing infrared heat energy. Averaged over the full depths of our oceans, the heating is equivalent to about 0.7 watts per square metre. That may not sound like a lot, but imagine 360 million square kilometres (139 million square miles) of ocean sprinkled with desk-spaced LED lights. In the forty-year period between 1980 and 2020, the heat content of the upper ocean (the first 2,000 metres, 6,560 ft) has been increasing at 10 worldwatts per year. In other words, the upper layer of ocean is like a giant kettle plugged into a power source ten times the size of the global economy. There is a relationship between CO_2 concentrations and temperature. Limiting greenhouse gas emissions by switching away from fossil fuels (as well as reducing other greenhouse gas emissions) will result in less global heating.

One trillion tonnes and the Climateclock.net

In 2009, Oxford climate scientist Myles Allen and his colleagues published a groundbreaking paper in the scientific journal *Nature*. The United Nations Framework Convention on Climate Change, born at the Rio Earth Summit in 1992, had the rather vague goal of limiting global warming to 'safe levels'. It took until the early noughties for scientists and politicians to put a number to the idea of 'safe', which was agreed to be a limit of +2 °C (3.6 °F) above the pre-industrial temperature in 1850 for anthropogenic warming. Modelling how the earth's systems respond to global carbon emissions is a complex and uncertain science, but Professor Allen and his colleagues managed to express the goal in a neat way. They calculated that to have a reasonable chance of limiting global warming to +2 °C (3.6 °F), global historic and future emissions of carbon should be kept to a maximum of 1 trillion tonnes (3.67 trillion tonnes of carbon dioxide, adjusting for molecular mass).

In other words, there is a very well-defined limit to how much ancient fossilized sunlight we can extract and burn. We've gone through about 70 per cent of our emissions budget already. We are currently emitting about 40 billion tonnes of carbon dioxide per year directly from burning fossil fuels (and a bit from producing cement). The Climateclock.net, a countdown clock set up by Professor Allen and his team, answers the question: given the current rate of emissions and level of human-induced warming, and assuming the emissions trend over the past five years continues into the future, how long will it be before the remaining allowable emissions for a rise of 1.5 °C (2.7 °F) are used up? At the time of writing, the answer is about twelve years, with us hitting 2 °C (3.6 °F) by 2052.

> **Currently, oil use in the developed world averages fourteen barrels per person per year. In the developing world, it is only three barrels per person. How will the world cope when billions of people go from three barrels to six barrels per person?**
>
> DANIEL YERGIN, *The Quest: Energy, Security and the Remaking of the Modern World* (2011)

After these dates, the temperature continues to increase further, assuming that current levels of emissions continue. If we want to *stabilize* the global temperature increase at 1.5 °C (2.7 °F), we've got to hit net zero global emissions by 2050. When we talk about net zero emissions, we're referring to a situation in which the total additions of greenhouse gases to the atmosphere are less than the sum total of those that we remove from the atmosphere – for example, by planting trees or using yet-to-be-invented large-scale carbon dioxide vacuums connected to some long-term storage system. To stabilize at a 2 °C (3.6 °F) increase gives us another twenty years before we need to reach net zero by 2070.

GLOBAL PRIMARY ENERGY CONSUMPTION IN 2019	
	% global primary energy consumption
Coal	26
Oil	31
Gas	23
Modern renewables	10
Nuclear	5
Traditional biomass	5
Total	**100**

Note the way we measure primary energy is complicated; we'll come back to that in Chapter 9.

We have known about climate change and the problem of burning fossil fuels for about forty years. Yet, during that time, our consumption of coal, oil and gas has risen steadily. In 2019, we broke the records for our consumption of oil and gas, and coal consumption remains close to its all-time high.

Humans have never used as much energy as we do now. There are currently roughly 8 billion people living on the planet, more of us than ever before, and our numbers are growing by about 1 billion every twelve years. This makes the fact that we need to go from 40 billion tonnes of carbon dioxide per year to zero (or less) by around 2050 even more challenging. To compound the issue further, the International Energy Agency predicts that if we carry on as we are, by 2050 a world population of around 10 billion will require 2 worldwatts – i.e. just under double the primary energy that we consume today.

We currently get just under 500 EJ (exajoules) from fossil fuels, 67 EJ from renewables and 25 EJ from nuclear. We cannot rely on nuclear or hydropower alone to replace fossil fuels. There are some serious limits to the expansion of large nuclear and hydropower schemes. Therefore, in just two to three decades, we need to increase our production of renewable energy by about ten- to twenty-fold overall.

There is no question that renewable energy is the future. There is no alternative. While it is very encouraging that some cities, nations and companies are committing to 100 per cent renewable energy futures, the fact remains that the whole world needs to become more or less 100 per cent renewable, as it effectively was in the pre-industrial era, and very soon.

04 WE WILL THRIVE ON SUNLIGHT

'It is entirely possible for the industrial nations of the world to terminate their dependence on non-renewable sources of energy and to create a gentler, fairer, more ecologically conscious civilization based on the indefinitely sustainable energies of the sun.'

GODFREY BOYLE (1975)

If the earth wasn't spinning, a day would last a year. Luckily, it spins on its own axis once every twenty-four hours on its 365-day orbit around the sun. The axis of the earth's rotation is not too far off something approaching perpendicular (it in fact wobbles over a 41,000-year cycle somewhere between around 22 and 25 degrees) to the plane of the earth's orbit. This makes the equatorial regions hot and the polar regions cold.

An enormous amount of solar energy hits the earth's surface each day. From the sun's perspective, the earth is a tiny flat disc 150 million kilometres (93 million miles) away. As we saw in Chapter 2, the solar flux (concentrated sunlight) hitting this disc is about 1,370 W/m². From the earth's perspective, the average solar flux is 342 W/m² at the top of the atmosphere (four times less since the surface area of the globe is four times the area of its silhouette). About 50 per cent of this energy bounces around until it eventually goes back into space thanks to clouds and the greenhouse effect. This leave us with on average 170 W/m² hitting the earth's surface.

This is a good figure if we are investigating the earth's overall radiation budget, but if we want to know more about the potential to capture solar energy, we need to start to look in more detail at the actual geography of sunlight.

Imagine a metre-by-metre-square solar collector that we can place anywhere on the earth's surface. Wherever we place it (apart from certain months at the poles) it experiences a sunrise, midday and sunset. The concentration of sunlight hitting the horizontal collector varies over the course of the day from zero in the night-time to its maximum around noon, before falling once again towards zero after sunset. The lower the sun in the sky, the greater the angle of incidence (measured from the vertical) and the less overall sunlight it captures. The solar collector will receive half the sunlight when the sun is at 60 degrees (measured from the vertical) compared with when the sunbeam is striking it directly overhead. The closer the collector is to the equator, the higher the daily path of the sun in the sky will be, and therefore, on average, the greater the amount of solar radiation the collector receives over the course of the day. This all assumes there are no pesky clouds getting in the way of the collecting. Clouds are a major factor that will affect how much sunlight the collector receives.

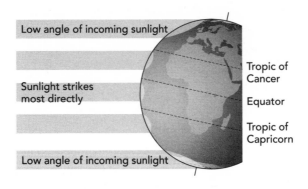

Let's return to global averages for a moment. Over the course of a year, 170 W/m² is about 1,500 kWh per square metre (170 × 365 × 24 watt hours) or about 4 kWh per day. This is a global theoretical average. According to the World Bank's Solar Atlas, solar radiation varies from over 2,000 kWh per square metre per year or more at the equator to a half or less of 1,000 kWh/m²/year in Northern Europe. Thanks to special solar radiation meters called pyranometers, we have some detailed maps of average solar potential at the global, regional and local scales. Weather systems (clouds in particular) play an important role in the solar potential of a particular region.

THE SUNNIEST PLACE ON EARTH

The sunniest places on Earth have, on average, fewer clouds and therefore less rain. A map of the world's hot deserts shows that they all intersect the 30 degrees north or south latitudes. This is no accident. The arrival of so much sunlight in the equatorial regions sets up giant, stable and predictable weather patterns that meteorologists call Hadley cells. Warm, moist air rises at the equator, creating tropical weather systems such as trade winds and thunderstorms. These travel north or south until the hot, dry air sinks as high pressure at the 30 degrees north and south latitudes. Yuma in Arizona, USA, is one of the sunniest places in the world. Daylight hours range from eleven hours in winter to thirteen in summer and on average nine days out of ten are cloud-free. Yuma has an average monthly Global Horizontal Irradiance (GHI) of 5.83 $kWh/m^2/day$, which is the total irradiance received by a square metre of surface horizontal on the ground, the sum of two types of sunbeam that arrive on a horizontal surface.

Two different ways to measure sunlight can give rise to some heated arguments about which regions are the sunniest in the world. The first is Direct Normal Irradiance (DNI), which measures the vertical component of radiation coming directly from the sun. The second is Diffuse Horizontal Irradiance (DHI), which measures sunlight that has been scattered by the atmosphere, sometimes called sky radiation. Most solar collectors are tilted at an angle towards the sun and some adjust and track the sun's daily path to maximize the amount of solar energy they collect. When a surface is angled, it receives both direct and scattered sunlight, but also a third component – sunlight reflected from the ground directly to the collector. PV (photovoltaic) systems mounted at a fixed angle towards the sun would use a third measure, Global Tilted Irradiance (GTI), which can be calculated from GHI and DNI. DNI is more relevant to sun-tracking solar collectors, since this more closely measures what they 'see'.

> **I see solar becoming the new king of the world's electricity markets.**
>
> FATIH BIROL, Executive Director, IEA (2020)

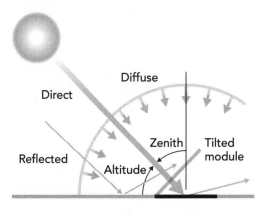

Which is the sunniest continent? Africa wins on one measure and Australia on another. These differences would be imperceptible to humans, but make a significant difference to what a solar collector on average 'sees'. The differences matter for solar farmers calculating how much energy, and therefore money, they will collect in a year.

	kWh/m²/day	
	GHI	DNI
Africa	5.71	6.25
Australia	5.59	6.59
South America	4.90	4.79
Oceania	4.19	4.41
Asia	3.81	4.43
North America	3.24	4.03
Europe	2.93	3.25

Solar thermal technologies

Technologies to collect solar thermal energy have been developed over a long period. Glass is one of the oldest, but still one of the most important solar energy-collecting technologies. It has the special property of being able to let daylight in, but trap some of the infrared solar radiation that might otherwise escape. A window is therefore a solar collector. Daylight streaming through a window is a form of passive solar technology. It's passive in the sense that it has no moving parts or pipes and relies on daylight to heat the air and surfaces behind it. In temperate or colder climates, humans and animals rarely miss a chance to passively heat themselves as they lounge in the sun. Buildings, too, can be designed with the passive collection of solar energy in mind – and this can dramatically reduce heating requirements over the course of year.

> **Even though solar is still a tiny amount of global energy, we are in the exponential phase now. It's going to be a wild ride in the years ahead.**
>
> JEREMY LEGGETT, Director of Solarcentury (2018)

Solar hot water heaters can take several forms. A black bag or hosepipe filled with water will heat up quickly on a sunny day. Fine if you are camping, perhaps – it's simple and effective – but poor at storing the hot water for the evening or next morning. The first patent for an active solar hot water collector was taken out in 1909 by William Bailey in California. He marketed it through the Day and Night Solar Water Heater Company and incorporated an insulated tank for storage. A century later, while the designs have become more sophisticated, the basic physics hasn't changed.

The technology is used in most countries and climates around the world. In Cyprus and Israel, for example, 90 per cent of homes have solar hot water systems thanks to government regulation and backing.

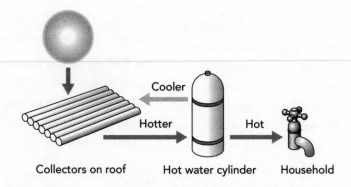

Cooler

Hotter

Hot

Collectors on roof   Hot water cylinder   Household

There are various types of solar collectors. The most basic use hot water as the circulation fluid in the collector and the tank. These are direct or open-loop designs. When the water is hot enough you simply open a tap. If it circulates (thermosyphons) naturally, this is called a passive system. If there is a circulation pump, it is called an active system. Most modern solar collectors are indirect or closed-loop systems that use a working fluid such as antifreeze in a separate loop that flows through a heat exchanger to transfer the heat energy from the collector to the hot water in the storage tank. These are much more efficient, especially if the working fluid is pressurized. Flat-plate collector designs are large copper surfaces usually enclosed in a glass-insulated box. The glass is an important feature – it increases the efficiency greatly, though there is still a considerable amount of heat loss. Evacuated tube collectors are designed to minimize the heat loss further using a vacuum to prevent convection heat losses, and are generally more efficient over a broad range of temperatures, cloud conditions and seasons.

The International Energy Agency (IEA) estimates that in 2019 about 684 million square metres (7.36 billion square feet) of solar collectors were installed globally. This provides about 480 GW (thermal) with an estimated output of 390 TWh – saving 42 million tonnes (46 million US tons) of oil and 135 million tonnes (149 million US tons) of CO_2. There are large markets for solar thermal collectors in China, USA, Germany, Australia, Mexico, Turkey, Denmark, Cyprus, South Africa and Greece. Solar water heating could power 15 to 30 per cent of hot water needs by the middle of the century, with significant economic fuel savings. Payback periods for solar hot water systems are as short as two to four years, depending on the technology and location.

Photovoltaic power

When light hits certain materials, it can create electricity. The photovoltaic effect, which is the flow of electrons excited by light, was first observed by Edmond Becquerel, a French physicist, in 1839. Willoughby Smith, a British telegraph cable engineer and scientist, is credited with discovering photoconductivity in 1873, which is the emission of electrons into a conductor. He accidentally discovered that his selenium test rods became more conductive when exposed to sunlight. In 1877, William Adams and Richard

Day observed the photovoltaic effect in solidified selenium and published a paper on the selenium cell.

The world's first solar PV cell is credited to the American Charles Fritts, who in 1883 constructed the first selenium solar cell. It had an efficiency of less than 1 per cent and it wasn't until 1954 that the first practical silicon solar cell was invented by Bell Labs with about 6 per cent efficiency. Bell Labs placed adverts at the time promising a source of 'limitless energy of the sun'. In 1958 the US satellite Vanguard 1 became the first to have solar electric power. It remains the oldest human-made object still in orbit!

Solar PV cells are made from thin layers of semi-conducting material – usually silicon – and are then connected together in various ways to form a panel or film. A lot of research continues to go into making the efficiencies of solar PV panels higher. The world record stands at 47 per cent for a system developed by the National Renewable Energy Laboratory in the USA in 2019, but these are very expensive and are predominantly used in the space industry. Typical commercially available efficiencies are still in the range of 15 to 20 per cent.

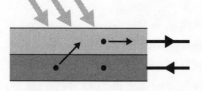

Solar PV is expanding at around 12 per cent per year. Total installed global capacity was 627 GW in 2019, compared to just 23 GW in 2009. Most of this is in China (about 200 GW), with the US, Japan, Germany and India making up 400 GW. Globally, the amount of electricity generated from PV is still relatively small – about 3 per cent – but in some countries it contributes much more, for example 10.7 per cent in Honduras, and over 7 per cent in Italy, Greece, Australia and Japan.

> **With this modern version of Apollo's chariot, the Bell scientists have harnessed enough of the sun's rays to power the transmission of voices over telephone wires.**
>
> *THE NEW YORK TIMES*
> (front page, 26 April 1954)

The cost of PV power has been declining rapidly and in some markets is already reasonably competitive with coal, gas and on- and offshore wind. Although the potential for distributed rooftop solar PV is huge, the vast majority of growth in PV is from large utility-scale, ground-mounted solar PV plants. These are becoming as big as the fossil-fuel plants in an effort to drive down costs. In 2019, thirty-five solar PV plants bigger than 200 MW were installed, with some as big as 500 MW (in Vietnam, Spain and China). India recently completed the two gigantic (over 2 GW) projects, Pavagada and Bhadla Solar Park.

The technical lifetime of these solar panels is increasing and is now about thirty years. This means that we are starting to see a substantial flow of broken and defective PV panels that require recycling. The total cumulative mass of installed PV panels in the world is currently around 5 million tonnes (5.5 million US tons), about the same as the total PV panel waste stream expected by the 2050s. PV panels are generally classified as general or industrial waste, and the EU, well aware of the coming wave of old PV panels, has set specific PV electronic waste recovery and recycling regulations and targets. About 75 per cent of PV panels sold today are silicon based (classed as mono-, poly- or multicrystalline), 11 per cent are thin film based (copper indium gallium selenide or cadmium telluride) and 14 per cent consist of other technologies such as organic PV cells and advanced crystalline silicon.

PV panels are mostly glass, along with polymers, aluminium, silicon, copper and silver, with traces of other metals such as zinc, nickel and tin. The solar PV industry and national governments are developing strategies to minimize the growing problem of PV waste using advanced design along with reduce, recycle and reuse strategies.

A 10 GW solar farm is being planned halfway between Alice Springs and Darwin in Australia. The Sun Cable project will export some of its electricity to Singapore via undersea HVDC cables. Some highly respected technology assessments, such as Project Drawdown, suggest a third or more of our global electricity needs will be met from distributed and utility-scale PV by 2050.

THE EVOLUTION OF THE WORLD'S LARGEST SOLAR PV FARMS

Year	Name	Country	MW	Area (km²/miles²)
2005	Bavaria Solarpark (Mühlhausen)	Germany	6.3	0.4/0.15
2006	Erlasee Solar Park	Germany	11.4	0.8/0.3
2008	Olmedilla Photovoltaic Park	Spain	60	2.8/1.1
2010	Sarnia Photovoltaic Power Plant	Canada	97	4.5/1.7
2011	Huanghe Hydropower Golmud Solar Park	China	200	5.6/2.2
2012	Agua Caliente Solar Project	United States	290	9.7/3.7
2014	Topaz Solar Farm	United States	550	19/7.3
2015	Longyangxia Dam Solar Park	China	850	27/10
2016	Tengger Desert Solar Park	China	1547	43/17
2019	Pavagada Solar Park	India	2050	53/20
2020	Bhadla Solar Park	India	2245	57/22

Concentrating solar power

By using mirrors to concentrate sunlight, it is possible to create very high temperatures to boil water and power steam turbines to drive electricity generators. There are four types of concentrating solar power (CSP) plants:

- Power towers consist of a large number of sun-tracking mirrors on the ground (called heliostats) all focusing on the top of a central tower that contains water, oil or salt, which is heated up to then drive a steam turbine.
- Parabolic trough concentrating collector systems focus the sun's rays into a pipe containing oil running along the focal point of a parabolic gutter-shaped trough. This can track the sun's path across the sky. Oil is heated to 200–400 °C (392–752 °F) and pumped away to generate steam/electricity.
- Parabolic dish concentrator systems place a small Stirling engine at the centre of a large radio-telescope-sized mirror.
- Linear Frensel collectors use flat mirrors to approximate the parabolic shape to produce superheated steam.

Most of the world's CSP is the trough type, though the power tower type is catching up and has the most favourable economic outlook. In 2019 there were just over 6 GW of CSP installed around the world. Spain is the leader with over 2.3 GW installed followed by the US with about 1.7 GW. Most recent CSP schemes involve some form of heat storage, as this is a relatively cheap way to store solar energy for around seven to ten hours. This enables CSP schemes to collect solar energy in the day and have it available to be converted to electricity for the evening peak in demand. Most CSP schemes store thermal energy in a giant cylindrical metal tank, similar to the large fuel tanks you might see at ports or airports. Molten salts are stored at around 565 °C (1,049 °F) inside the tank and amazingly lose only one degree of heat per day, meaning that heat can be stored for weeks. The tanks last about thirty years before they need to be repaired, but the salts can be used over and over again.

CSP is very well suited to regions with DNIs above 2,500 kWh per metre squared per year of sunlight radiation. This includes many regions, such as the deserts of Australia, China, India, Pakistan, Russia, the southern United States, Central and South America, Africa (north and south), the Mediterranean, the Middle East, and Iran. It would seem a reasonably conservative estimate that CSP could power up to 10 per cent of the world's electricity needs by 2050.

Solar fuels

The future is electric. It is also likely to involve the production of hydrogen gas from renewable sources of

> ❝ **PV plus electrolysis is a pretty clean, green way to make hydrogen.** ❞
>
> DAME MARY ARCHER (2018)

energy. Hydrogen is an extremely useful companion fuel to renewable electricity and without doubt one of the most exciting technologies in our solar energy future. The good news is that there is a lot of hydrogen on Earth. It mostly likes to combine with oxygen in the form of water, plants and animals, and in fossil fuels. Only a tiny fraction of hydrogen gas exists in the atmosphere, as it has a tendency to escape the earth's gravity into space.

Renewable electricity can be used to split water into hydrogen and oxygen via electrolysis. The prospect of generating 'green hydrogen' from sunlight is real and is a critical part of net zero-carbon scenarios for the future (more on this in Chapter 9).

As we saw in Chapter 2, solar power represents by far the greatest promise for utilizing renewable energy. The vast potential for direct solar energy in the future is calculated by assuming that a small fraction of total land area (less than 2 per cent) could be used for centralized PV power plants. The potential for CSP overlaps with some of this land – those sunny areas with a very high level of solar irradiance. The potential for solar heating is so vast that it is really only limited by the demand for heat itself. Not all regions of the world have high solar potentials, but many do. Not all regions of the world are self-sufficient in fossil or nuclear fuels, either. The prospects for harvesting and exporting solar energy either directly using low-loss direct current transmission grids or indirectly through hydrogen pipelines are truly enormous.

GLOBAL DIRECT SOLAR ENERGY POTENTIAL BY REGION (WORLDWATTS*)

	Min	Max
Middle East and North Africa	0.7	18.5
Sub-Saharan Africa	0.6	15.9
Former Soviet Union	0.3	14.4
North America	0.3	12.4
Central Asia	0.2	6.9
Latin America and Caribbean	0.2	5.7
Pacific OECD	0.1	3.8
South Asia	0.1	2.2
Pacific Asia	0.1	1.7
Western Europe	0.0	1.5
Central and Eastern Europe	0.0	0.3
TOTAL	**2.6**	**83.2**

* 1 worldwatt is 599 EJ per year, which is our global primary energy consumption in 2019

The IEA's Sustainable Development Scenario shows a twelve-fold growth in the electricity generated from PV by 2040 compared to 2018 and a sixty-seven-fold increase (albeit from a relatively small baseline) in electricity generated from concentrated solar power. In 2020, the IEA confirmed that solar is now the cheapest electricity in history. In the coming decades, in one form or another, humanity can and will thrive more and more on direct energy from the sun.

05 CUBISM IS THE ART OF WIND ENERGY

'The cost of electricity for onshore wind is already competitive compared to all fossil fuel generation sources and is set to decline further.'

INTERNATIONAL RENEWABLE ENERGY AGENCY (2019)

Wind has powered the geographies of human civilizations for millennia. As we saw in Chapter 3, there is evidence of sail boats travelling on the Nile around 5000 BC, and we also know of crab claw sails used on the first ocean-faring boats that travelled around the islands of the Indian and Pacific Oceans between 3000 and 1500 BC. We have a long history of transforming wind energy into mechanical power for water pumps, flour mills and sailing ships. Many micro inventions using springs, rods and shutters have helped windmills cope with different high-wind conditions and to self-regulate. Not least was the improvement on spring and roller reefing sails in 1813 by the English engineer and inventor William Cubitt, whose patent sails replaced the commonly used cloth sails with shutters.

The work of the devil

The first wind turbine to produce electricity was built in 1887 by the Scottish electrical engineer Professor James Blyth. He used a cloth-sailed turbine to charge batteries that in turn were used to light his home in Aberdeenshire, Scotland. In a moderate wind, his designs could power ten twenty-five-watt bulbs, but his offer to use surplus electricity to light the main street of Marykirk, where he lived, was rejected, as the townsfolk thought electricity was 'the work of the Devil'.

One of the earliest larger electric wind turbines was built by US inventor Charles Brush in 1888 in Cleveland, Ohio. The 12 kW multi-bladed 17-metre (56 ft) diameter design (imagine a sandcastle wind toy, rather than today's modern propeller designs) incorporated the first step-up gearbox, which increased the revolutions per minute from the wind rotors to the electric motor. Brush used it to charge a bank of batteries to power his mansion.

With remarkable foresight, in 1891 the Danish scientist Poul la Cour (nicknamed the 'Danish Edison') used a wind turbine to produce electricity to power an electrolyzer, which in turn produced hydrogen and oxygen that was used to power a school's gas lamps. Today, wind-fuelled electrolysis to turn water into hydrogen as a way of storing

> **When the winds of change blow, some people build walls and others build windmills.**
>
> CHINESE PROVERB

wind energy in low-carbon electricity systems is one of the most exciting areas of advanced renewable energy systems. Storing wind energy can in turn help manage electricity grids. We call it the 'green hydrogen' revolution.

The evolution of wind power over the twentieth century has been driven primarily by Denmark and the United States. In the 1980s, Germany and Spain also joined the party. More recently, several Chinese companies have entered the market. Wind is sometimes referred to as *indirect* solar energy. The winds are driven directly by the sun's energy, which perpetually creates blooms of relatively short-lived low- and high-pressure systems. These are the isobar contours we see in the weather forecast. The atmosphere is a restless mass of air mixing through thermal convection. In the equatorial regions, warm moist air rises to about 18 kilometres (11 miles) and then moves either north or south. Some sinks at 30 degrees latitude and some at around 60 degrees latitude where it meets cold polar air. The rising and sinking of air creates the predicable large-scale patterns of trade winds that sailors have used for millennia. There are also local wind systems set up by land masses, mountain systems and oceans.

The global wind energy atlas

Estimates of the approximate size of the global wind energy resource increase over time as new turbine designs get bigger, cheaper, stronger and capable of being sited offshore at greater depths. Technically, there is more than enough wind to supply up to six times current global electricity production. The global atlas of wind energy resource is not distributed evenly across the land and coastal regions and is mainly concentrated at latitudes 30 to 60 degrees north and south of the equator. Not all windy places on Earth are suitable sites for large wind power projects, and geography and economics (distance to cities, state of the electricity grids, social acceptance) will play a major role in shaping how much of the global wind energy resource is eventually developed.

The global technical potential for both onshore

> As yet, the wind is an untamed, unharnessed force, and quite possibly one of the greatest discoveries hereafter to be made will be the taming and harnessing of it.
>
> ABRAHAM LINCOLN
> (1860)

and offshore wind is around double our global electricity requirements by 2050 or around 80,000 TWh/year. In practice, wind energy is predicted to supply about a quarter of our overall global electricity requirements by 2050. Over half the global total additional growth of onshore and offshore wind energy in the next three decades will be in Asia. Most of it will be onshore and in China, with North America and Europe making up the majority of the remainder. Over half the electricity used by Denmark is supplied from wind and over a third in Lithuania and Ireland, with several other nations not far behind.

Turbine technologies

Modern wind turbines are highly sophisticated devices incorporating cutting-edge scientific and technological expertise, computational design, and operation and maintenance procedures. They consist of a hub that holds the rotor blades, which is mounted onto a box called the nacelle, and a tower. Wind turbine blades are very similar to aeroplane wings – they are made of composite materials such as glass or carbon-fibre-reinforced plastics and designed to bend and flex.

The nacelle contains drive shafts, a gearbox, a generator, brakes, and pitch and yaw motors. The rotational mechanical energy of the turning hub is transferred to the gearbox through the drive shaft, where the rotational speed is increased to power the electrical generator. The yaw

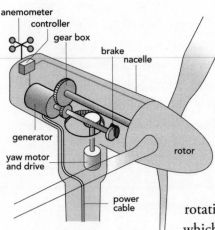

anemometer
controller
gear box
brake
nacelle
generator
yaw motor
and drive
rotor
power
cable

drive helps the turbine to swivel horizontally to ensure that the rotors are always facing into the wind. Optimizing the pitch or angle of the blades relative to the wind is another key component technology – in a similar way to the rotating blades of a helicopter, which work to increase or decrease vertical acceleration, the turbine blades swivel in their sockets. The pitch and brake systems regulate the hub speed to maximize energy production as well as protecting the turbine from damage in high winds.

Small innovations continue to increase the overall efficiency of catching the wind. A simple recent example is vortex generators. These are small plastic pads stuck to the root base of turbines that alter the air flow and can give a 1 to 3 per cent improvement in energy output from a wind farm. In addition to the wind turbine itself, a wind farm needs a transformer and connections to a nearby electricity grid. Transformers are vast, costly metal machines, especially for offshore wind farms. An 800 MW AC to DC converter, for example, weighs about 10,000 tonnes (11,000 US tons).

Kinetic energy

Wind is a stream of moving air. At low speeds we don't normally feel it, but the air isn't as thin as we think. It weighs about 1.23 kg per m^3 (2.7 lbs per 35 ft^3) at sea level. Putting a hand outside a car window, we can feel the force of the wind increase as the car speed (wind speed) increases. The wind power increases as the cube of the wind speed. This means, for example, that the power doubles when the wind speed increases from 4 to 5 metres per second. The cube law (see the box on the next page) is important, as it tells us that if we want to get more power from the wind, we can change two things: we can make the turbine blades bigger (increasing their swept area) or we can put the wind turbines in windier places (increasing wind velocity).

Doubling the size of the area of the rotor blades will double the power output. But with the same-size rotor blades we can still more than double the power output by shifting from a wind site with an average wind speed of say 6 metres (20 ft) per second to one that has an average of 8 metres (26 ft) per second. Maximizing the power output from a given size of wind turbine lowers the cost of the electricity generated. The wind power cube law is the reason why some of the earliest wind turbines in the UK, for example, were sited in windy places, on hills and often visible for miles. The economics of wind power is all about increasing the size of the rotor blades and then finding the greatest wind. So it is perhaps no surprise that the latest wind turbines

THE CUBE LAW

Wind power designers, engineers and entrepreneurs always have the cube law in the forefront of their minds. Wind power is an example of the conversion of kinetic energy of the wind into mechanical rotational energy to drive the magnets in an electric motor to produce electricity. The kinetic energy (joules) of a moving object with mass m (kg) travelling at speed v (m/s) is calculated using the formula:

$$\text{kinetic energy} = \tfrac{1}{2}mv^2$$

The mass of the cylinder of air passing through the wind turbine blades per second, m (kg), is itself proportional to the wind speed, v (metres per second), the area of the wind turbine rotor, A (m²), and the density of air, ρ (1.23 kg per m³, 2.7 lbs per 35ft³ at sea level).

$$m = \rho Av \text{ (equation 2)}$$

Combining these two equations gives us the famous cube law result for the power in the wind, P (watts):

$$P = 0.5\rho Av^3 \text{ (equation 3)}$$

This same formula can be used for the power of a tidal, ocean or river current, too, though the value of ρ will be different (see Chapter 6).

are enormous compared to their predecessors, and are being put in windier places.

Of course, wind speeds are often greater offshore (for the same hub height) but the costs of installing as well as operating and maintaining wind farms at sea are much greater. The costs of offshore wind are currently about double what they are onshore.

> **It feels like there's a new generation of turbines introduced almost every year now.**
>
> MICHAEL RUCKER, founder and CEO of Scout Clean Energy (2019)

Wind turbines stay asleep until they reach their 'cut-in' speed of around 3–4 m/s (10–13 ft/s). They then start turning and the power output increases with wind speed until the turbine reaches its 'rated power level', often in the region 11–15 m/s (36–49 ft/s). As the wind increases further, the turbine's control mechanisms (stall and pitch control) prevent the blades from moving faster and overloading the turbine's mechanics and electronics. Most modern turbines stop producing energy at windspeeds above 25 m/s (82 ft/s) (the 'cut-out' speed) to prevent damage. There is a theoretical upper limit of just under 60 per cent for the amount of energy that can be extracted from the wind. This is known as the Lanchester–Betz limit.

Onshore versus offshore

Windmills on the land have been around a long time. There were perhaps 200,000 windmills at their peak in Europe in

1850 (compared to perhaps 500,000 waterwheels) and 10 per cent of those in the Netherlands are still standing! Yet offshore wind farms are relatively new. The wind offshore of a coastline is often stronger and smoother in its flow due to less friction and disruption from the landform. The first offshore wind turbine was the Vindeby Offshore Wind Farm on the Danish island of Lolland, completed in 1991. It was decommissioned in 2017 after twenty-five years, which is a typical lifetime for a wind turbine due to stress fatigues. Since then, there has been rapid growth and interest in offshore wind energy. By 2019, the total amount of wind power globally had reached 650 GW and has been growing at nearly 10 per cent a year. Around 30 GW of this is offshore wind, and over half of the growth is currently in China.

The maximum depth for piling wind turbines offshore is about 40 metres (130 ft), due to the maximum size of jack-up cranes. For a long time, this has meant that whatever wind resource lies outside the 40 m (130 ft) coastal contour on sea charts was out of bounds. Most offshore wind farms to date are therefore relatively close to shore. The North Sea is particularly shallow and home to the world's largest offshore wind farm, completed in 2020. Hornsea One lies 120 km (75 miles) off the Yorkshire coast and supplies enough energy to power 1 million UK homes. It consists of 174 × 7 MW turbines standing on 100-metre (330 ft) tall towers. The swept area of each blade is larger than the

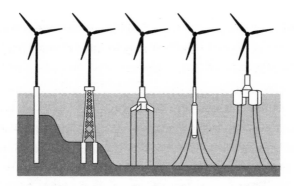

London Eye. Although 7 MW is large, it is only about half the size of the current large-scale turbines on order.

There is much excitement in the wind energy industry at the prospect of floating wind turbine platforms that could mean that wind farms could be placed at 100–200 metre (330–650 ft) ocean depths. Floating turbines dramatically expand the amount of coastal area where offshore wind could be sited. The technologies for securing the floating platforms to the seabed are familiar to the offshore oil and gas industries, which are increasingly investing in the development of wind energy.

Improving efficiencies

Over the twenty-or-so-year lifetime of a wind turbine currently operating, the amount of greenhouse gases caused by its manufacture are very low compared with its energy output. Turbines usually pay back the energy needed to manufacture and install them in around three to nine months. Better wind turbines, higher hub heights, longer and lighter blades and larger swept areas will all make wind energy more efficient and cheaper. The size of onshore turbines is growing rapidly – the average size has doubled in the last seven years. The capacity factor for a wind farm is the proportion of the time it runs at its maximum rated output. Many small innovations continue to improve wind farms, and capacity factors for onshore will rise from 35 per cent to close to 50 per cent by 2050. For offshore, from 43 per cent to nearly 60 per cent.

The cost of wind energy has fallen by a third in the last decade because of myriad design and technology improvements, from the blades to the rotors, motors and transformers. The cost of onshore wind electricity is competitive with natural-gas electricity generation and is set to decrease further in the coming decades. The cost of offshore wind is more than onshore. About two-thirds of the cost of onshore wind is the turbine itself while for offshore wind farms, the grid connection, civil works and increased operation and maintenance costs make up roughly 50 per cent of overall costs. The growth of wind power has been driven by government energy subsidies called feed-in tariffs, but increasingly wind energy projects are competing with fossil-fuel electricity supply options in the energy market.

Your own pigs don't stink.

DANISH WIND
FARMER PROVERB

THE WORLD'S LARGEST TURBINES

Siemens Gamesa have begun production of their 14 MW SG 14-222 DD offshore wind turbine, currently the largest-capacity wind turbine in the world, and the company plan to start wide-scale rapid production in 2024. The turbine has 108-metre (354 ft) long blades on an overall 222-metre (728 ft) diameter rotor, with a swept area of 39,000 m^2 (420,000 ft^2) and a power output of up to 15 MW. Remarkably, a wind farm using these models will be able to produce 25 per cent more energy than its predecessor. It is a 'Class 1' wind turbine designed to perform in the windiest conditions (average wind speed of 10 m/s or 36 kph or 22 mph). Turbine classes are determined by three parameters – the average wind speed, the maximum expected fifty-year gust, and turbulence. Turbulence wears out the component parts of a wind turbine. The SG 14-222 DD is able to operate in wind speeds that would normally mean other turbines would need to shut off; one turbine produces enough electricity to power approximately 18,000 average European households every year. Put another way, it can power two average UK households for a whole day in a single rotation of its blades.

There have been many critics of the wind power industry over the years. In some countries, this wasn't helped by the first wind farms being sited in places of outstanding natural beauty and ownership being centralized. In Germany and Denmark, they were owned by their communities. When communities own wind farms they are less conscious of the visual intrusion and noise that wind turbines can bring.

Growing pains in wind power

Collecting and converting any source of energy itself takes energy. Energy has to be invested in order to get a return flow of energy out. We call this the energy return on energy invested ratio (EROEI). In the case of wind farms, there is energy embodied in the steel towers, the concrete, the manufacture of the blades and the electronics in the nacelle. As wind turbines have become larger and moved offshore, the EROEI ratio for wind over a twenty-year lifetime has declined from around 20:1 in the early 2000s to as low as 15:1 in recent years for some offshore wind farms.

Another issue in the wind power industry is what to do with old wind-turbine components, in particular the composite blades, once they run their course. The earliest fleets of wind turbines are now reaching their retirement age and we are starting to think about what to do with them. To meet our climate targets with wind power within the next couple of decades, we will need to replace the equivalent of the entire stock of global wind turbines every year, each year. This is on top of increasing the global stock of wind turbines. The wind turbine blade industry currently makes over 80,000 blades per year and uses more than 750,000 tons of composites, which is 90 per cent of total global production of technical composites. This will grow ten-fold by the middle of the century.

There is also the issue of rain. A summer rainstorm can create significant problems for turbine blades, which can literally take a hammering from these large, hard and fast droplets. The larger the rotor, the higher the speed at the tip of the blade. The 50-metre (164 ft) rotor blades of the 1990s typically had a tip speed of 65 m/s (234 kmh, 145 mph). The tip speed goes up with rotor diameter, so for an 80-metre (262 ft) blade length, the speed is >70m/s (252 kmh, 157 mph), for a 100-metre (328 ft) blade length >80 m/s (288 kmh, 179 mph) and for a 160-metre (525 ft) blade length >90 m/s (324 kmh, 201 mph). The energy impact of the raindrop is related to the square of the tip speed, so the hammering impact on the leading edge of the blade has doubled as the blade length has doubled, creating new design challenges for blade manufacturers.

It's easy to manipulate energy statistics to make the current contribution of renewable energy look small. This is especially so when we use primary energy statistics (Chapter 9). The share of renewables and of wind in overall primary energy consumption can appear surprisingly small (as nuclear power did in the 1950s). However, it is clear that the future is mostly electric and this should be the yardstick we use to measure progress. One number should shock and amaze us: wind currently produces 6 per cent of global electricity output, which is just 2 per cent of the global technical potential we have tapped into so far. Floating offshore windfarms are not by any means the technical frontier for wind, with high-flying wind kites and airborne wind drones now being researched.

In 2020, the share of renewables in global electricity production was about 28 per cent. Various future energy scenarios envisage wind contributing about a third of future global electricity production – in the region of 8,000 TWh by 2040. To meet that goal, the wind industry needs to be six to ten times its current size, installing something like 500 GW per year while maintaining 10,000 GW capacity. And let's not forget that this also means replacing an additional 500 GW per year while adding that further 500 GW per year. It's a big ask.

06 PLANTS AND WATER ARE POWERFUL

'The planet needs trees. If there is indeed that carbon dioxide out there in the atmosphere, the only species on the planet that can actually trap it for us in a natural process of photosynthesis are the trees.'

WANGARI MAATHAI (2009)

Humans are solar powered. We live on plants that photosynthesize sunlight, and so too do the animals we eat. Bioenergy along with waterpower are two of the oldest forms of renewable energy that humans have harnessed.

Bioenergy is as old as the trees. Actually, it is a lot older. Put simply, bioenergy is energy sourced and converted from biomass. In the world of renewable energy, biomass means organic materials (compounds that contain carbon and hydrogen bonds), such as wood and other farmed energy crops, agricultural and food-system wastes.

All the world's biomass lies in the biosphere – the roughly 10-km (6 mile) layer above and below sea level that contains all the living and recently dead plants and

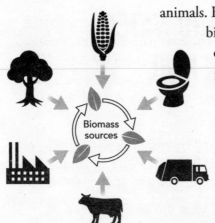

animals. Estimating the amount of biomass in the biosphere is one of those 'how many grains of sand are there?' questions – it's very hard to quantify. Our best and rather rough guess is that there is about 550 GtC (606 US gigatons C) distributed among all the kingdoms of life. Plants, which are mainly on land, make up 450 GtC (496 US gigatons C) – the vast majority. Animals account for about 2 GtC (2.2 US gigatons C) and are primarily marine based. Bacteria (about 70 GtC, 77 US gigatons C) and archaea (microorganisms, about 7 GtC, 7.7 US gigatons C) are mainly located below ground.

The increase in human biomass has gone hand in hand with a seven-fold decline in wild mammal biomass. The biomass of livestock (about 0.1 GtC, mainly cattle and pigs) is about twenty times that of the total mass of 8 billion humans (about 0.06 GtC), which in turn is about nine times greater than the biomass of wild mammals (about 0.007 GtC).

Around 2.1 billion years ago, bacteria began transforming the earth's atmosphere. In particular they started to raise the levels of oxygen from almost zero to the present level of about 20 per cent (by volume). Bacteria's oxygen production also created stratospheric ozone, which was critical in shielding the earth from high-energy UV radiation and allowing the evolution of more complex plants and animals like us to take place.

Photosynthesis and plants (ingested directly or via animals, recent or ancient) drive everything, including the human economy. Only a tiny fraction of the biomass in the biosphere – for example, wood, straw and dung – is readily available for human use. From a renewable energy perspective, biomass is the continual replenishment of the store of chemical energy in the biosphere through the process of photosynthesis – the process in which plants use solar energy to convert carbon dioxide and water into carbohydrates and at the same time release oxygen as a by-product. This is the famous photosynthesis equation that you might remember from school, which shows how plants convert six molecules of carbon dioxide and water using sunlight into one molecule of plant sugar and six molecules of oxygen:

$$6CO_2 + 6H_2O + \text{solar energy} \rightarrow C_6H_{12}O_6 + 6O_2$$

From this equation we can see that plants need carbon dioxide. Plants, animals and their ecosystems are a significant component of the natural carbon cycle. Indeed, plants are playing a critical role in shielding us from the worst effects of anthropogenic climate change. We saw in Chapter 3 that plants are responsible for sequestering half of all the fossil carbon we have emitted since the start of the Industrial Revolution. A very useful service. In fact, rates of global surface temperature increase would be double or more without them.

Modern humans have been around for about 220,000 years. For 99 per cent of this time we have relied on bioenergy – mainly in the form of food, wood and animal wastes. It may shock you to know that one in ten of us still cooks on open fires today. In 2018, 2.8 billion people did not have access to clean energy for cooking, relying on wood, coal, charcoal or animal waste to heat their food.

> **There are three types of biomimicry – one is copying form and shape, another is copying a process, like photosynthesis in a leaf, and the third is mimicking at an ecosystem's level, like building a nature-inspired city.**
>
> JANINE BENYUS (2011)

But unsustainable cooking methods using wood fuel and cattle dung contribute to 4 million extra household pollution-related deaths per year in low- and middle-income countries, with emissions linked to a range of heart and lung diseases, and cancer, to name a few. Some particulates from this household air pollution, in turn, also contribute to climate change.

THE UN SUSTAINABLE DEVELOPMENT GOAL 7

In 2015, the United Nations adopted a set of Sustainable Development Goals (SDGs) as a universal call to action to end poverty, protect the planet and ensure that all people enjoy peace and prosperity by 2030. Goal 7 is about ensuring access to affordable, reliable, sustainable and modern energy. Each goal has specific associated targets. The targets for goal 7 are:

7.1 By 2030, ensure universal access to affordable, reliable and modern energy services

7.2 By 2030, increase substantially the share of renewable energy in the global energy mix

7.3 By 2030, double the global rate of improvement in energy efficiency

7.A By 2030, enhance international cooperation to facilitate access to clean energy research and technology, including renewable energy, energy efficiency and advanced and cleaner fossil-fuel technology, and promote investment in energy infrastructure and clean energy technology

7.B By 2030, expand infrastructure and upgrade technology for supplying modern and sustainable energy services for all in developing countries, in particular least developed countries, small island developing states, and landlocked developing countries, in accordance with their respective programmes of support

The world of bioenergy

Bioenergy comes in many forms – solid, liquid, gas, traditional and modern – and has many different uses. Correctly and accurately describing how much of our renewable energy comes from bioenergy is a particularly tricky and slippery task. About 10 per cent of our global energy consumption in 2018 was bioenergy (traditional and modern). This is about 70 per cent of all the renewable energy we used in 2018. The amount of traditional biomass used in the global final energy demand is over 6 per cent – a staggering amount. The modern bioenergy contribution to global final energy demand is nearly three times higher than wind and solar PV combined. It is mostly used for heating, although the use of bioenergy for electricity and transport biofuels has been growing rapidly in recent years.

The biomass that is used as a feedstock to produce bioenergy can either be classified as primary or secondary. Primary biomass is directly derived from plants and can include:

- Sugary biomass – sugarcane, sugar beet sorghum (sweet)
- Starchy biomass – wheat, maize (corn) and cassava
- Cellulosic biomass – miscanthus, switchgrass, short rotation willow, eucalyptus, seaweeds
- Woody biomass – wood fuels

- Oily biomass – oilseed rape, soybean, palm oil and jatropha

Secondary biomass sources include:

- Forest residues from tree felling made into wood chips
- Wood residues in the pulp and paper industry
- Wheat straw
- Bagasse – sugar cane residue
- Rice husks
- Animal wastes
- Sewage sludge
- Municipal solid waste
- Commercial and industrial biomass wastes (e.g. oils, tyres)

Different feedstocks can be treated thermally or bio-chemically to produced different energy sources and/or fuels.

By comparison, today's industrial roundwood (logs) production is 0.03 worldwatts, while the global harvest of major crops (cereals, oil crops, sugar crops, roots, tubers and pulses) is about 0.1 worldwatt. The market potential to increase land-based biomass is 0.4 worldwatts.

GLOBAL TECHNICAL POTENTIAL FOR LAND-BASED BIOMASS SUPPLY BY 2050

Biomass category	2050 technical potential (worldwatts)
Residues from agriculture	0–0.1
Biomass production on surplus agricultural land	0–1.2
Biomass production on marginal agricultural land	0–0.2
Forest biomass (primary wood and secondary residues)	0–0.2
Dung	0-0.1
Food waste and discarded paper/wood packaging	0-0.1
Total Technical Potential	**0.1–1.7**
Total Market Potential	**0-0.7, with a best estimate of 0.4**

More than half of modern bioenergy is used for heating, mostly in the industrial and agricultural sectors, about a fifth is used for transport fuels, and 10 per cent for electricity production, which is the fastest-growing of them all. Bio-heat can also be used to produce electricity through co-generation of power and heat. Biomass feedstock industries such as paper and board, food sector and wood-based industries often use their residues to supply bio-heat and electricity. Major areas of activity in modern biomass include biomass pellets, liquid biofuels (ethanol, biodiesel) and biogas. In the IEA's Sustainable Development Scenario, global transport biofuel production triples to 300 Mtoe (Millions of tonnes of oil equivalent) per year by 2030, and bioenergy used in power generation doubles from around 600 TWh to 1,200 TWh.

Negative emissions technology

When plants, or their derivatives, grow, they absorb carbon dioxide and when they are burned, they release the carbon back to the atmosphere. If the carbon dioxide that is produced from the combustion of plants could be captured and stored – for example, underground – then biomass-based energy production offers the opportunity to reduce atmospheric concentrations of CO_2 (negative emissions). This is sometimes called bioenergy with carbon capture and storage, or BECCS. Bioenergy is the only renewable energy that offers this prospect: wind turbines and PV farms cannot directly help to capture carbon dioxide. At the end of 2019, there was one large-scale project

> **The use of plant oil as fuel may seem insignificant today. But such products can in time become just as important as kerosene and these coal-tar products of today.**
>
> RUDOLF DIESEL (1912)

capturing and geologically storing CO_2 from a corn ethanol plant in Illinois, USA.

Controversies and problems

The amount of additional biomass that can be produced without causing negative social and economic impacts through competition with land and water for food production is uncertain and highly controversial. One of the main criticisms is that land dedicated to growing crops for fuel (for example, corn to produce ethanol as a transport fuel) is not available for food production. Another is that for some primary or secondary biomass sources, the overall net energy gained from conversion is limited, or even negative.

Generally, woody energy crops have a fuel–energy output to input ratio of around 10–20:1, whereas ethanol from grain can be as low as 1:1. Proper lifecycle analysis is a complex modelling task and is open to manipulation from a pro- or anti-bioenergy perspective. It must consider the different primary or secondary bioenergy sources, the conversion technologies and the final energy uses, and take into account what might have happened with the feedstocks in the baseline. Amazingly, half of all bioenergy remains the traditional use of wood and dung burning on traditional stoves. While governments work on making modern, cleaner forms of bioenergy and electricity available, any short-term improvements that could be made to the design and use of traditional cooking and heating practices would be highly cost-effective.

Hydropower

Water is a remarkably eccentric liquid. Over the aeons, H_2O's peculiar thermodynamic qualities have, in no small measure, helped shape the earth's climate system that we know and love. Water is good at storing heat, moving heat in liquid form and grabbing heat from one place and dumping it in another. It's a busy part of the earth's thermodynamic balancing cycle.

The fact that water has a relatively high boiling point (100 °C or 212 °F) compared to the average surface temperature of the earth (about 15 °C or 59 °F) means that a lot of it is able to sit around in liquid form. Hence 70 per cent of the earth's surface area is covered with an average of 4,000 m (13,120 ft) of water. That's a lot of water. The specific heat capacity of water that so enthralled James Joule is higher than most common substances – about three times greater than soils and rocks, for example – and is why it plays an important role in our climate system.

The 'wateriness' of water is down to the relatively strong hydrogen bonds between molecules in its liquid state. Water molecules are relatively social – they enjoy sticking together. It takes five times more heat to evaporate a quantity of water than it does to raise it to boiling point (from 0–100 °C or 32–212 °F at atmospheric pressure). If you think about it,

> **People are moving away from the fossil fuel-based economy to a more renewable economy. That is what is called the "transition-town" movement. There are 300 towns in Britain that are making this transition. Taking energy from solar power, from wind power, from waterpower.**
>
> SATISH KUMAR (2014)

a pan of water quickly boils compared to the amount of energy it takes to evaporate the water in it altogether. And yet water also has a relatively low viscosity, which means it will readily self-organize streams of warm and cold currents in the oceans, transporting huge quantities of energy towards the poles.

About 40 PW of the sun's daily flux of energy on the earth is used to drive the earth's water cycle. This is about 2,100 times the earth's total human primary power demand. It is all put to good use powering the water cycle, lifting water from the oceans and the land up into the atmosphere, where it circulates as water vapour before condensing into clouds and falling as rain. The amount of water that is evaporated (averaged over the earth's surface) is 1 metre (3.3 ft) per year (about 3 mm, or 0.1 inches per day). About 85 per cent of global evaporation is from the oceans.

Evaporation is related to the net radiation at the surface – so it is greater towards the equator and over the ocean. Around 80 per cent of all the water vapour falls

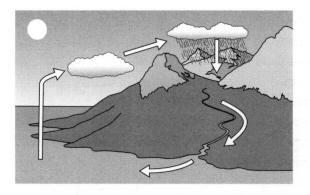

back directly onto the oceans and 20 per cent falls on the land. Sixty per cent of all the precipitation on land is re-evaporated, about 10 per cent runs off into the oceans in a disorganized fashion and about 30 per cent is carried by rivers.

Hydropower is an old energy technology that has become a significant part of global electricity production and will remain so into the future. The waterwheel was possibly the first non-human/non-animal source of kinetic to mechanical energy technology. The first waterwheels (sometimes called Greek or Norse wheels) were relatively inefficient horizontal types. There is some evidence that vertical water wheels were in use in Ptolemaic Alexandria and during the Roman empire, and water mills spread rapidly in Europe during the medieval period for milling and mechanical power. By the time of the Domesday Book in 1086, there were over 5,000 water mills in south-east England.

HOW MUCH HYDROPOWER IS POSSIBLE?

We can do a quick back-of-the-envelope estimate of how much hydropower – in theory – might be available. The total land-based run-off and river discharge is about 50,000 km³ (12,000 miles³) per year. The average height of continental land is 850 metres (2,800 ft) above sea level. This immediately gives us a figure for the global potential energy of all this rain:

Potential energy (J) = mass (kg) × gravity (m/s^{-2}) × height (m)
= $(50,000 × 10^{12})$ × (10) × (850)
= $425 × 10^{18}$ J

Recalling our mnemonic from Chapter 2: **M**illions of **G**reen **T**urbines **P**ower **E**lectrons (mega-giga-tera-peta-exa), this equals 425 exajoules.

This is roughly 0.8 worldwatts. However, only about 10 per cent of this is in reality potentially available once practical considerations such as geography, agriculture and engineering are taken into account. However, 8 per cent of the overall global energy requirement in the form of electricity is an extremely important contribution.

Hydropower is already the dominant source of electricity for some regions and nations. About 16 per cent of global electricity generation in 2019 was hydropower, the third largest source of electricity after coal at 39 per cent and natural gas at 24 per cent. It supplies 95 per cent of Norway's power, and over half of all power in Brazil, Canada and the South and Central Americas. China produces three times more hydropower than its nearest rival, Brazil. China's annual production of 1,270 TWh is equivalent to the total electricity generation of Central and South America, or France and Germany combined, and just short of India's current total electricity production.

Streams, rivers and lakes can come in all shapes and sizes and in different geographical terrains, and so different turbine technologies have evolved to make use of varying scenarios. Hydropower engineers often describe the world of hydropower in two dimensions: water pressure (proportional to the height or 'head of water') and the water flow. High heads are usually above 100 metres (328 ft) and low heads around 10 metres (33 ft) or less. The higher the head, the less flow is needed for the same power output.

> ❝ It should be borne in mind that electrical energy obtained by harnessing a waterfall is probably fifty times more effective than fuel energy. ❞
>
> NIKOLA TESLA (1905)

Water pressure increases by around 1 standard atmosphere (atm) every 10 metres (33 ft).

Modern hydro turbines come in two basic technological forms: impulse turbines, which convert kinetic energy into mechanical energy by changing the direction of a flow of water, and reaction turbines, which convert kinetic into mechanical energy by stepping down the water pressure. Impulse turbines have jets of water hitting wheels with paddles and buckets on them. They often resemble a wheel with a hundred ice-cream scoops closely spooned together, and the ice-cream scoops are partly cut away to allow the water to escape. A jet of water hits one side of the scoop, then tracks around the scoop, changing direction by almost 180 degrees and then leaving by a cut-away part of the scoop. The water jets propel the wheel. The jets are surrounded by air with no significant air pressure changes. Turgo wheels are a variant of the Pelton wheel and resemble some sort of fossil shell creature.

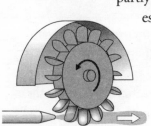

Impulse turbine

In reaction turbines, water arrives at the turbine under pressure and is deflected by runner blades and vanes back out along the axis of the turbine at a lower pressure. The pressure drop is a key part of the energy transfer. An example of these, Francis turbines, named after the

British-American civil engineer James B. Francis, are installed in the majority of the world's hydro schemes. They can operate in a range of environments from low- to high-head schemes. The world's largest hydropower plant, China's Three Gorges Dam on the Yangtze River, uses thirty-two 700 MW Francis turbines weighing 6,000 tonnes (6,600 US tons) each. Propellers can also be used as turbines, such as axial-flow turbines, that are used in tidal barrages schemes. If the angle of the blades can be changed, they are called Kaplan turbines.

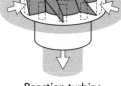

Reaction turbine

Sustainable energy scenarios suggest the share of hydropower in overall electricity generation may increase by 50 per cent by 2040 to 21 per cent of total global electricity generation.

Debates about energy policy often rapidly deteriorate into simplistic pros and cons. We see this with almost all forms of energy (coal, oil, nuclear), including the different renewable energy sources. Biomass energy and hydropower are no different. Not all land is suitable for biomass bioenergy production, but some certainly is. Not all rivers can be sustainably dammed, but they can come close. Hydropower and bioenergy already comprise a significant chunk of global energy production. There is no doubt that for certain countries and ecosystems, hydropower and properly managed biomass harvesting for energy production can be sustainable energy technologies for the future.

07 THERE'S PLENTY OF BOUNTY BELOW

'Science, my lad, has been built upon many errors; but they are errors which it was good to fall into, for they led to the truth.'

JULES VERNE, *Journey to the Centre of the Earth* (1871)

Ninety-nine per cent of planet Earth is hotter than 1,000 °C (1,832 °F). The earth is, in fact, a giant leaky heat battery. About half of the 40 TW of heat flux coming from the earth's core is primordial and the other is due to radioactivity. The earth's primordial geothermal energy can be tapped directly as a source of heat (as the Greeks and Romans enjoyed in their geothermal spas) or it can be used to produce steam to drive steam turbine electricity generators, replacing natural gas, coal, biomass or nuclear power plants.

The first prototype geothermal electricity plant came on stream on 4 July 1904 in Larderello, Italy, when Prince Piero Ginori Conti, the head of a boric acid company, managed to light five light-bulbs from a dynamo attached to a reciprocating steam engine driven by geothermal heat.

A year later he increased power production to 20 kW and by 1916 it distributed 2,750 kW of electricity around 30 km (19 miles) to the cities of Pomarance and Volterra. Larderello now produces about 10 per cent of the world's entire supply of geothermal electricity, enough to power around a million Italian households.

The ability of steam and water mixtures to power generators or heat exchangers depends on the various temperatures and pressures, and the dryness and wetness of the steam. Just as Inuit can distinguish many different kinds of snow, a good steam turbine engineer can differentiate between several types of steam.

The geothermal heat map

It is a misconception that countries need to have geysers spewing at the surface to have some geothermal potential. There is considerable potential resting quietly beneath much of the planet.

Notable geothermal power nations are clustered around the Pacific Ring of Fire, a huge ring in the basin of the Pacific Ocean characterized by active and dormant volcanoes. A range of countries and regions make direct use of geothermal heat for district heating and agricultural purposes, including China, the USA, Sweden, France, Turkey, Japan, Norway and Iceland.

The technical potential of geothermal electric power is between 100 and 1,100 EJ/yr (assuming a depth of

up to 3–10 km, 2–6 miles). This is up to 2 worldwatts. Direct thermal extraction is in the range of 10 to 312 EJ/yr (up to about 0.5 worldwatts). As we saw in Chapter 2, in terms of technical feasibility, the global geothermal potential is probably closer to 1 worldwatt. This heat is truly renewable in the sense that it will be replenished over the long term by the earth's natural heat flux. On a global basis, these figures might at first seem implausible, but many national studies have concluded that there is significant untapped geothermal potential. Currently, the top ten nations for geothermal electric power capacities are: USA, Indonesia, Philippines, Turkey, New Zealand, Mexico, Kenya, Italy, Iceland and Japan.

The earth's temperature increases with depth. Near the surface (the earth's crust), the rate of warming, or geothermal gradient, is 15–30 °C (27–54 °F) per km (0.6 miles). Different kinds of rocks have different sorts of conductivities depending on their chemical and physical composition as well as their porosity. Different zones have different thermal

> **Geothermal will be a game changer for Eden, Cornwall and the UK.**
>
> SIR TIM SMIT, co-founder of the Eden Project (2020)

conductivities and therefore heat gradients. At the base of the crust the temperature is approximately 1,000 °C (1,832 °F) and this increases to 6,000 °C (10,800 °F) at the earth's core.

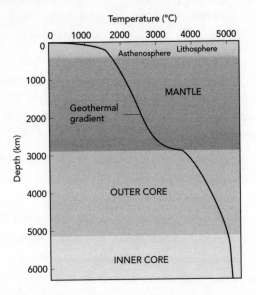

The top 10–15 metres (33–49 ft) of ground is heated by solar radiation and acts as a seasonal fluctuating heat store in temperate climates. Heat pumps (fridges working in reverse by collecting heat from the 'outside' and dumping it 'inside') can transfer heat to or from the ground. They can be connected to different sorts of heat exchangers ranging from shallow (1–2 metre or 3.3–6.6 ft depth) horizontal networks of pipes to deeper (typically up to 250 metre or 820 ft) borehole pipes. Technically, shallow geothermal heat pumps do not produce geothermal energy because they exploit the sun's heat energy as it warms the surface.

Geothermal energy is anything collected at depths typically up to 200–300 metres (660–980 ft). The normal lifespan for a deep geothermal well is around thirty years, or until the surrounding rocks have cooled down too much from the cold water injected into the well, which generates the steam. However, if left for twenty-five to thirty years, the geothermal well will have heated up again.

Oil companies commonly drill down to 5,000 metres (16,400 ft), where temperatures reach 170 °C (338 °F), with some test wells as deep as 12 km (7.5 miles). However, deeper drilling becomes a challenge due to the high temperatures. Plastics, electronics and steel all start to develop problems beyond about 200 °C (392 °F). With advances in new materials and clever engineering, the geothermal industry hopes to get down to temperatures of 300 °C (572 °F) within a decade and 500 °C (932 °F) after that.

Geothermal technologies

Power generation requires geothermal fields of medium to high heat content and there are five types of technologies in use or under development today.

Direct dry steam plants use low-pressure, high-volume fluid produced from a steam field of 150 °C (302 °F) or higher, which is at least 99.995 per cent dry to avoid scaling and erosion of the turbine or tubes. Direct dry steam plants commonly use condensing turbines and range in size from 8 MW to 140 MW.

Flash plants use steam at temperatures greater than 180 °C (356 °F) obtained from a separation process called flashing. These are the most common type of geothermal electricity plants in operation today.

Binary plants use steam from low- or medium-temperature geothermal fields (100–170 °C or 212–338 °F), which enters a heat exchanger and in turn heats a process fluid in a closed loop. The process fluids are ammonia and water mixtures or hydrocarbons, and have lower boiling and condensation points that better match the geothermal resource temperature.

Hybrid plants are standalone geothermal power plants with an additional heat source, for example heat from a concentrating solar power (CSP) plant added into the process. The additional heat is added to the geothermal brine, increasing the temperature and power output.

Enhanced Geothermal Systems (EGS) are engineered reservoirs made where there is hot rock but not enough natural permeability or fluid saturation. They use fracking techniques to cause pre-existing fractures to re-open, creating enhanced permeability that allows fluid to circulate and the transport of heat to the surface where electricity can then be generated. While advanced EGS technologies are young and still under development, EGS has been successfully realized on a pilot scale in Europe. The largest EGS project in the world is a 25 MW demonstration plant in Cooper Basin, Australia, with an estimated potential to generate 5,000–10,000 MW. Realizing the upper estimates of geothermal global technical potential will mean developing commercial-scale EGS technology. Like hydropower, geothermal power projects are capital-intensive, but have low and predictable operating costs.

Messing about with hot water, steam and toxic gases under high pressure from wells that are 0.5–4 km (0.3-2.5 miles) deep is asking for all kinds of physical and chemical trouble. Geothermal energy may be renewable, but it is not without its environmental issues. The main problems can include site noise and water pollution in the exploration and development phases, ground subsidence, gaseous pollution and induced seismicity.

The IPCC estimates that the full lifecycle CO_2-equivalent emissions for geothermal energy technologies are moderate compared to other renewable energy sources. EGS power plants hold the prospect of possibly being designed with zero direct emissions if they are able to capture and store carbon.

Ocean energy

The oceans are a gigantic store of untapped heat and kinetic energy. They contain vast amounts of thermal energy and are currently absorbing the majority of the additional warming caused by anthropogenic climate change in their upper layers.

The oceans trap both wind energy and heat energy from intense tropical sunshine. The theoretical potential for ocean energy is very likely to exceed our global primary energy use and covers six distinct technologies, including:

- Wave energy converters – wind energy transferred and stored temporarily in ocean waves and collected by snake and duck-like devices
- Tidal range schemes – water is made to flow through turbines generating energy from the rising and falling tide
- Tidal and ocean currents turbines – chunky-looking propellers put on the seabed or suspended below the surface to make use of tidal currents, which can be 5–12 knots in UK waters, for example.
- Ocean thermal energy conversion (OTEC) devices – these exploit the temperature differences between warm surface waters and the cooler deeper ocean
- Salinity gradients devices – these harness the power of osmosis to create a salt battery

Ocean technologies are all still in the early stages of development. The IPCC notes that the theoretical potential for ocean energy technologies has been estimated at as much as 7,400 EJ/yr (about 15 worldwatts), though this is highly uncertain. What is certain is that there is a whole world out there, down there, that in comparison with solar and wind power, is relatively unexplored and could provide a sizeable chunk of our energy requirements in the future.

Wave power

Wave energy is a form of concentrated and stored solar energy, because waves are driven by the winds, which are, in turn, driven by sunshine.

The power of a wave is proportional to the square of its wave height, so a 3-metre wave is nine times more powerful than a 1-metre wave. The power is measured by their energy density per metre along their wave fronts. About 95 per cent of the energy of a wave is contained in the surface layer, which is a quarter of the wave's wavelength deep. The average North Atlantic waves are about 60 kW/m, but this can rise to 1,700 kW/m in a storm. The ocean's ability to concentrate wind energy into wave power can be spectacular. On 7 February 1933, the USS *Ramapo* oil tanker encountered the tallest rogue wave ever recorded in the Pacific. It measured 34 metres (112 ft) in height, which translates to a power of 7.7 MW/m!

> **There are no quick fixes to long-term energy challenges. To find solutions, governments and industry benefit from sharing resources and accelerating results.**
>
> ANA BRITO E MELO, Ocean Energy Systems Executive Secretary (2017)

'Wave energy converters' still come in many shapes and sizes – a sure sign that a technology hasn't yet reached maturity. The first patent for a wave energy converter dates back to 1799, filed in Paris by Pierre-Simon Girard, a French mathematician and fluid mechanics engineer. In 1910 the French inventor Bochaux-Praceique dug a vertical borehole in a cliff near Bordeaux to drive a turbine, supplying power and lighting to his house. Similar shore-based oscillating water-column schemes remain a key area of development today.

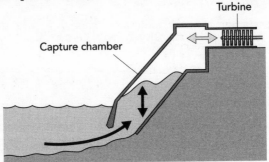

The IPCC estimates that the technical potential for wave power is 115 EJ/year or about 0.2 worldwatts. Like wind power, wave power is intermittent and therefore not suitable as the baseload of an electricity system, with most of the wave resource between latitudes of 30 to 60 degrees.

Currently there are around twenty wave energy converters at the megawatt scale and some 10 MW scale under construction. By 2040, second generation systems could reach the 2–10 GW scale. The European wave power industry would like to supply 10 per cent of Europe's energy by 2050 (about 190 GW).

The engineering challenges of working in high-energy offshore environments means wave energy remains a very high-cost source of renewable energy, though the environmental impacts of wave power are possibly less than other forms of renewable energy.

Tidal range and currents

Tidal energy is predictable and therefore potentially a source of baseload electricity. The IPCC estimates that the world's theoretical tidal power potential (tidal range plus tidal currents) is in the range of 3 TW, with 1 TW located in relatively shallow waters. However, only a fraction of this potential is likely to be exploited. One study estimated that the potential of the twenty-eight best sites globally amounted to 360 GW (about ten times the Three Gorges schemes), which, while not significant globally, could be transformative for some coastal regions. Currently, 90 per cent of the world's tidal barrage capacity is accounted for by two projects: the 240 MW La Rance station in France and the 254 MW Sihwa plant in the Republic of Korea.

Tidal barrages are not without their environmental implications, in particular for coastal ecosystems and habitats, and this may be one reason why we haven't seen the

> **The sea itself offers a perennial source of power hitherto almost unapplied. The tides, twice in each day, raise a vast mass of water, which might be made available for driving machinery.**
>
> CHARLES BABBAGE (1832)

development of more large-scale commercial schemes. There has been much more interest in tidal current turbines, which are easier to deploy and have much lower ecological impacts.

The global total power of ocean currents is surprisingly small. It has been estimated at 100 GW (a tiny fraction of the earth's current 19,000 GW primary energy consumption). Some major currents, such as the Gulf Stream between the Bahamas and Florida, are responsible for as much as 20 GW (twenty nuclear power plants, or one of the Three Gorges Hydro dams working flat out).

To extract energy from tidal flows, the technologies need to be reversible as the tide flows in and out. If the flow is in a river or a steady one-way ocean current, the technologies can be simpler in design.

There are three principal types of tidal, ocean and river current technologies:

- Axial-flow turbines (look like common propeller-type wind turbines)
- Crossflow turbines (tall vertical spindles)

- Reciprocating devices (they look a bit like the nodding donkeys/pumpjacks that cover the US oil belts)

Many areas around the world have potential usable ocean currents, and tidal stream devices are rapidly evolving towards one design in particular: the horizontal axis turbine mounted on the seabed or suspended under a floating platform.

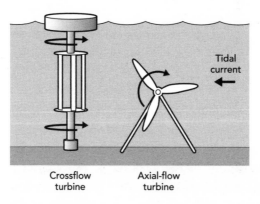

Crossflow turbine   Axial-flow turbine

The power of flowing tide with water density ρ (kg per cubic metre) with velocity (v, m/s) with a cross section area (A metres^2) is:

$$P = 0.5\rho Av^3$$

It is the same formula for flowing air (Chapter 5). The density of air is about 1.23 kg per m^3 compared to a value for ρ of 1,025 kg per m^3 (2,260 lbs per 35 ft^3) for seawater. In other words, for the same flow speed, water offers 837 times more energy per metre. A tidal current of 3 knots has the same energy density as a steady wind stream at 29 knots (a fair old blow). A tidal flow of 3 knots is a reasonably common current for UK waters, where currents can be as much as 12 knots. An average wind speed of 140 knots, which is a catastrophic category 4 hurricane on the Saffir–Simpson Hurricane Wind Scale, is needed to match the power density of a 12-knot current.

Ocean thermal energy conversion

Where there is a temperature gradient, there is a flow of heat. The heat flow can be put to useful work. Sea surface temperatures in the earth's tropical oceans can exceed 25 °C (77 °F), with temperatures 1 km (0.6 miles) below the surface being between 5 and 10 °C (41 and 50 °F). A temperature difference of 20 °C (36 °F) is enough to power a steam turbine driven by either seawater or ammonia (in a closed loop so that it doesn't kill the fish!), or a mixture of the two. The warm seawater is used to produce a vapour that acts as a working fluid to move turbines. Although interest in ocean thermal energy conversion (OTEC) started as early as the 1880s, the first OTEC plant was built in Cuba in 1930 by French industrialist and inventor, Georges Claude, also known as the 'Edison of France'. It generated up to 22 kW before being destroyed in a storm.

A handful of OTEC projects have been built or are in the pipeline. The largest to date was a 1 MW plant in Hawaii, now decommissioned. In the US, China and Martinique, some ten MW plants are in development. The technology represents the lion's share of the estimated technical potential of all ocean energy – and this is despite it being one of the most immature of all ocean technologies.

> **A new era of ocean exploration can yield discoveries that will help inform everything from critical medical advances to sustainable forms of energy.**
>
> PHILIPPE COUSTEAU, JR, oceanographer and activist (2012)

Osmotic power

Where there is a ready supply of both fresh and salty water, such as at the mouth of a river, there are some exciting and rapidly developing technologies that can harness the chemical potential between the two water sources. Using osmosis to mix freshwater and seawater across a semi-permeable membrane releases energy as heat that can be captured as pressure and then converted into useful energy forms. Osmotic power could be generated globally wherever there are river estuaries to play with.

The potential of salinity gradient power is considerable. The energy released from 1 m³ (35 ft³) of fresh water with 1 m³ (35 ft³) of seawater is equivalent to the 2 m³ (70 ft³) of water falling from a height of around 8 metres (26 ft) (an amount that hydro engineers would happily play with). The energy release can be as much as ten-fold using saturated brines (e.g. from waste water treatment or desalination plants). When salty water is fed between a pair of ion-exchange membranes with fresh water on the other side, the salt ions diffuse from the salty water into the fresh water. The positive ions will diffuse through one membrane and the negative through another, which creates positively and negatively charged poles, similar to a battery.

The IPCC have estimated the technical potential for osmotic power generation as 1,650 TWh/yr (6 EJ/yr). The International Renewable Energy Agency (IRENA) estimates the total technical potential for salinity gradient power to be around 647 gigawatts (GW) globally, or 19 per cent of electricity consumption in 2018. There are no large-scale demonstration projects currently, though there are several advances taking place in the laboratory.

Both geothermal and ocean energy technologies require massive investments in the next decades to achieve their potential. At first glance, both resources may seem extreme environments in which to operate, but it's worth bearing in mind that coal, oil and gas exploration and extraction present some equally extreme and similar environments. Most of the world's investments in renewable energy are in solar and wind energy, but with some more imagination and relatively small amounts of funding for research and development, geothermal and ocean sources could represent a sizeable chunk of tomorrow's sustainable energy economy.

08 UTOPIAS ARE ELECTRIC

'You never change things by fighting existing reality. To change something, build a new model that makes the existing model obsolete.'

BUCKMINSTER FULLER (1982)

Electricity is the future. Remarkably, this is as true today as it was for its pioneers Franklin, Faraday, Tesla and Edison in the eighteenth and nineteenth centuries. Our utopian dreams are electric: smart homes and cities, electric cars, electric aircraft and maybe even ships. The transition to a renewable-energy-powered electric future throws up some particular engineering and economic challenges.

Electricity is modernity's life blood. For 150 years the electricity transmission and distribution grids have evolved slowly and incrementally. All that is about to change. In many countries, wind power and solar PV power have become competitive with fossil fuels, even at a time when coal, oil and gas are modestly priced by historic standards, and regulators are paying attention to decarbonizing electricity networks in line with commitments under the Paris Agreement on climate change.

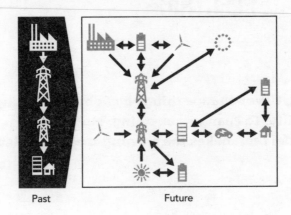

Past   Future

This all presents new challenges for managing the spatial and temporal geographies of supply and demand on our electricity grids. Even if, globally, the overall amount of future electricity demand didn't change much, simply coping with more renewables on the grid would be a significant engineering challenge. However, we are expecting a considerable increase in electricity consumption in the next three decades and this presents a double challenge. On top of that, nearly a billion of the poorest people on the planet remain unconnected to any kind of electricity grid at all. It is safe to say that our electricity grids are in for some drama as they evolve to be more renewable, diverse, distributed, flexible and interactive.

Electricity is power

The first electricity networks powered street lighting. In most modern economies, today's electricity networks power almost every aspect of our lifestyles from cooking to heating and cooling to the internet.

On 12 January 1882, three years after the invention of the carbon filament incandescent light-bulb, the world's first commercial coal-fired power station was commissioned. It was known as the Edison Electric Light Station, located at 57 Holborn Viaduct in London, and it powered 968 lamps. Edison opened a second light station in September that same year at Pearl Street Station in New York. He was using direct current (DC), so his power stations had to be near his customers to avoid large power losses. His rival, George Westinghouse, argued for alternating current (AC), which could be carried over longer distances with less loss, but this had the complication of needing to keep different power stations on the network precisely in time with each other to avoid damage. AC systems won and still dominate the main transmission and distribution grids in the world today.

As global technological revolutions go, the Holborn Light Station was an inauspicious start. Quickly running up massive financial losses, it closed in September 1886 and the lamps were converted back to gas. Such hiccups aside, myriad small-scale electricity grids were quickly established, and electricity became the oxygen of the modernization of industrial society.

> **With great power comes an even greater electricity bill.**
>
> UNKNOWN

It is hard to overstate the historic impact of the discovery of electricity on the world's economy and culture. Electricity took centre stage, for example, at the World's Columbian Exposition in Chicago in 1893. Just thirty years later, three-quarters of US households had been electrified. The political power of electricity was irresistible and intoxicating. It was clean at the point of use: no smoke, no fire, no dust, just lovely bright lights and endless ideas for electrical devices to replace manual labour tasks in the home, offices and industry. By the time of the Chicago World's Fair in 1933, electricity had become *the* symbol for modernity. It fuelled utopian visions of abundant cheap, clean energy to empower society and remove much day-to-day hard graft and toil.

Our twentieth-century electric transition depended heavily on the thermal combustion of fossil fuels. In 1900, less than 1 per cent of all fossil fuels consumed in the world were converted to electricity. Today, 64 per cent of coal and 40 per cent of natural gas is used to power our electricity grids.

Just as it was the case in the early days of urban town gas networks (and our present urban internet providers), lots of different companies competed to supply electricity on their own networks. However, this meant replication of generation assets and wires and increases in cost. The chaos, squabbles and skulduggery of decentralized private electric networks eventually gave way to the introduction of state-regulated electric utility monopolies. Electrifying economies was extremely capital-intensive and led governments all over the world to declare electricity a natural monopoly and therefore subject to regulation. Electric utility companies were born and with them the roots of the centralized electricity systems of the twentieth century. These are the same systems, more or less, that we have inherited today. Their characteristics are large, dependable nuclear and fossil-fuel power stations delivering electricity through passive distribution networks to unreactive electricity consumers.

In terms of network ownership and distribution, the pendulum has swung back to the private sector in the last few decades. However, the physical nature of how electricity grids are built, owned and managed has remained remarkably stable since the early days. Modern electricity systems have a combination of very powerful high-voltage supply generation, such as nuclear power stations or large gas-fired power stations, connected to an electricity grid.

We can think of the electricity grid as a network of major motorways connected to a smaller distribution network of trunk roads that connect major users, such as hospitals and factories, as well as smaller-scale, less powerful supply generation sources, such as wind farms, solar farms and battery systems, on lower-voltage distribution networks, and smaller roads that connect individual households.

Forecasting future electricity supply and demand is a notoriously difficult crystal-ball-gazing exercise. Most nations and international organizations now consider alternative future energy and electricity scenarios. There is usually a baseline scenario where current trends continue and a scenario that meets our global requirements for a sustainable energy system, often related to the 2 °C (3.6 °F) upper limit set under the Paris Agreement and the UN's Sustainable Development Goals.

GLOBAL ELECTRICITY SUPPLY BY SOURCE (%)

Source	2018	2040
Coal	38	6
Oil	3	1
Gas	23	15
Nuclear	10	12
Hydro	16	17
Wind and solar	7	41
Other renewables	3	10
(Subtotal all renewables)	26	68
Total	100	100

The IEA's 2040 Sustainable Development Scenario is one such image of the future. It shows global electricity demand up by 45 per cent, increasing from 27 to 39 TWh of electricity by 2040. Not *such* a dramatic forecast, you might think. However, a closer look reveals that renewable's share of global electricity rises from 26 per cent in 2018 to 68 per cent in 2040. If you are in the electricity business, this is very big news indeed. If you are in the renewables business, it is phenomenally, mind-bendingly exciting.

Decentralized, decarbonized and digitized

The UK National Grid Company talks about the future of electricity grids in terms of the three Ds: decentralized, decarbonized and digitized. The geography of electricity is changing. On the supply side it is increasingly about the geographies of the wind and sun. On the demand side it is about the geographies of new demands for electricity, such as electric vehicles, and the possibility of buildings, factories and car chargers that respond to the needs of the grid, using smart meters to turn themselves up and down, and on and off.

Our global electricity networks are already creaking. Even such advanced electric economies as those in Europe and North America regularly face the tangible risk of imbalance in supply and demand, and blackouts do occur. The grids in many developing countries are in a far worse state, with inadequate, underdeveloped or even non-existent electricity networks. The UN's Sustainable Development Goal 7 (see Chapter 6) aims to connect the remaining 800 million people on Earth to an electricity grid by 2030. Politicians euphemistically worry about 'keeping the lights on', but in truth, things are far more serious. Our lifestyles are intertwined with electricity: electricity powers the internet, which in turn powers the economy. At the same time, the internet is increasingly powering the smart electricity system. You can see the problem here!

Electricity networks that have more renewable electricity fed into them must be designed, built and operated to cope with greater fluctuations in the unevenness, or the potential mismatch, between the supply and demand on the grid at any moment. Wind and sunshine are less predictable than natural gas or nuclear power, though these traditional sources are not without their unexpected power downs.

New technologies, new business models and new rate-charging structures are all part of a rapidly evolving electricity landscape. The falling cost of renewable energy is a key driver of the transition to low-carbon electricity networks. In many countries, renewables (wind and solar) are the cheapest way of adding extra supply-side generating capacity in response to growing demand or the need to decommission older fossil or nuclear power plants. The cost of onshore and offshore wind as well as solar PV has been declining rapidly over the last decade. They are now cheaper than new fossil-fuel plants.

SUPPLY AND DEMAND

Electricity systems deliver electric power to consumers. Households measure power in kW, organizations usually in MW and nations in GW. Power is the rate of delivery of energy and in electricity systems the maximum rate is the capacity of the system. A nuclear power station has a capacity of about 1.5 GW, a large modern wind turbine 8 MW and an individual PV panel around 0.25 kW. The energy delivered by electricity systems is the amount of power (kW, MW, GW) multiplied by the time it is delivered (hours). Quantities of electricity delivered to customers are therefore measured in kWh, MWh or GWh. The load on an electricity system is the rate at which all the users on an electricity network demand energy from it. Loads can be measured at an instant (e.g. noon on Christmas Day) or as averages over weeks or months. Electricity is generated from a number of different resources on the supply side of the network. When customers on the network change their technologies or behaviour to influence when and how much electricity they need, we call these demand-side resources. Energy efficiency (e.g. LED light-bulbs), battery systems, smart meters and smart devices are all demand-side resources.

A second key driver of change is the Paris Climate Agreement to limit global warming to well below 2 °C (3.6 °F) above pre-industrial levels. All over the world, governments have mandated various decarbonization programmes to reach net zero by the middle of the century. China, for example, has announced its intention to be carbon neutral by 2060 – an amazing prospect given that it is currently the world's largest emitter of carbon dioxide and has a rapidly developing economy with electricity consumption growing at a whopping 8 per cent per year. The secret is that, while China uses a lot of coal, it is also a massive user and manufacturer of such renewable energy technologies as hydro and wind turbines, and solar PV panels. China's energy future (and ours, too) is largely in its hands.

A third driver is the global shift away from internal combustion engine vehicles to purely electric ones. Dozens of countries and major cities around the world have announced phase-outs and bans of internal combustion engine vehicles in favour of electric vehicles. In the future, millions of the latter will need charging from the network. This presents both a challenge and an opportunity in terms of managing the network load. China has just under half the world's stock of pure and hybrid electric vehicles (about 3.3 million out of a total of 7.2 million). Although electric vehicles currently account for only about 1 per cent of global car stock, they are growing exponentially at a phenomenal 40 per cent per annum. The global stock of electric vehicles (excluding two- and three-wheelers) is expected to reach 250 million by 2030.

As well as playing a crucial role in decarbonizing the transport sector, electricity will help decarbonize a considerable chunk of the heating and cooling sector through the use of air and ground-source heat pumps, as well as direct electric heating to help manage load on the network.

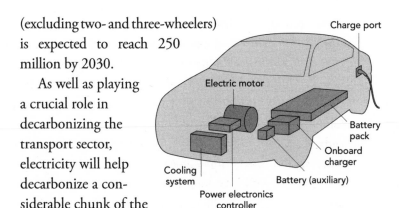

A balancing act

In the future, electricity networks will need to be more active and responsive to the weather, to consumers and to new technologies, such as electric vehicles and electric- and heat-storage devices. Electricity grids must cope with all extremes, such as the very cold, cloudy, windless winter's day with high surge demand as well as the very windy, sunny day with low demand. The grid has to get ready for changing daily load profiles, much greater overall flows, as well as greater extremes in peak flows between regions and national interconnectors.

A range of supply- and demand-side measures will help us manage the system. Energy devices are part of the

Internet of Things that are embedded with sensors and software that can exchange data autonomously over the internet. Greater numbers of smarter smart meters and sensors will transform how households and organizations use electricity and interact with the network system. Many more domestic, business and industrial electricity consumers will become energy suppliers or 'prosumers', who both consume and produce energy by installing their own solar PV, wind farms or battery bank.

Supergrids and microgrids

Supergrids are high-voltage transmission grids covering large geographical areas. A supergrid is part of the Desertec proposal to use parts of the deserts in North Africa to supply a large share of Europe's electricity needs. Supergrids can help areas the size of continents several thousand kilometres across take advantage of continental-size weather systems. Low-pressure depressions in the northern hemisphere are typically about 500 to 1,000 km (300–600 miles) wide and pass by in two days or so. Supergrids will shunt large quantities of electricity very long distances using low-loss, high-voltage direct current (HVDC) power lines (with losses as low as 2 per cent per 1,000 km, 600 miles). Of course, there are political dimensions to be overcome, just as is the case with today's electricity interconnectors and natural-gas grids.

At the other end of the scale we will see many more microgrids. These are a group of customers (loads) with distributed generation assets (wind farms, solar farms, gas power stations) that are managed as one distinct entity and are capable of being disconnected from the wider macro grid. They are also a key tool in the management of future electricity networks, allowing whole branches of the network tree to be safely switched off and back on when needed to balance the overall system.

> Solar power, wind power, the way forward is to collaborate with nature – it's the only way we are going to get to the other end of the twenty-first century.
>
> BJORK, *Dazed & Confused* (2011)

Storing electricity

As the amount of variable renewable electricity supply grows, the ability to store electricity as part of the overall management of the network becomes much more important.

Chemical batteries are expensive. An enormous amount of research and development money is pouring into lithium-ion and other exotic battery technologies, with some spectacular results. However, by far the most common large-scale power-storage technology in the world today, pumped-storage hydropower, has been used for

> We haven't solved all the problems inherent to this chemistry, but our results do show routes forward.

CLARE GREY, Professor of Chemistry, University of Cambridge (2015)

decades as a way of storing electricity and accounts for 96 per cent of installed global electricity grid storage. Water is pumped uphill at times of low demand and then released later to power turbines at times of high demand. It is a relatively cheap form of electricity storage and is surprisingly efficient (the full cycle is about 90 per cent energy efficient). A few concentrated solar power plants can store heat energy ready to generate electricity and there are a handful of mechanical flywheel projects (about 1 GW worldwide). There are also some compressed air and even gravity storage systems. In the future, new fleets of electric vehicles parked overnight could become another mass source of electricity storage and supply.

Storing wind and solar electricity as green hydrogen

Storing electrons is tricky and expensive. Converting them to hydrogen gas and storing that could be a much cheaper option. Instead of moving electrons in cables, the renewable energy future may be about the large-scale production of hydrogen from solar and wind electricity, and its storage and transmission through pipelines. Large-scale storage of hydrogen will help balance future energy demands with the variable supply from the sun and wind. The IEA's 2040 Sustainable Development Scenario sees hydrogen consumption increase from close to zero in 2018 to the equivalent of around 15 per cent of 2018 global hydropower production by 2040. This may be a small fraction of the overall energy consumption, but it is expected to increase rapidly beyond the middle of the century.

For a long time, hydrogen has been touted as an alternative to internal combustion in the transport sector, for example in the form of hydrogen fuel cells. This remains the case, but there is renewed interest in the use of hydrogen in other sectors, such as industry, domestic energy and power generation.

There are several industrial ways to make hydrogen. But there is only one way to make hydrogen without creating carbon dioxide as a pesky by-product. We call this method 'green hydrogen' to distinguish it from 'blue hydrogen' (made from natural gas with carbon capture and storage) and 'grey hydrogen' (made from natural gas with the carbon dioxide released into the atmosphere). Green hydrogen production uses renewable electricity to electrolyze water into hydrogen and oxygen.

In some cases, our existing fossil-fuel infrastructure can

> **Renewable hydrogen will play a major role in the next phase of the energy transition.**
>
> ISABELLE KOCHER (2018)

be used in the transition to sustainable energy. Hydrogen can be mixed in with natural gas at up to 20 per cent (by volume) and sent through the existing gas distribution grids with minimal modifications to grid infrastructure or to domestic end-user appliances. However, beware – your cooking flame may change colour. In Europe, there are projects exploring how hydrogen generated from North Sea wind farms, for example, can be stored in the many gas fields that exist there.

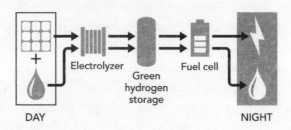

Our electricity grids are facing the perfect storm. Imminent blackouts, climate change, variability from renewables, electric vehicles and smart grids are a heady cocktail of risks and opportunities to manage. They require truly historic levels of investment and innovation to rise to these challenges.

When energy is cheap, the sun is shining and the birds are singing, and we waste energy like there is no tomorrow. It is much easier to carry on wasting it than to think about smart ways to avoid using it in the first place. When prices rise, the rain clouds come and darkness falls, we then, of course, have more appetite for such innovative thinking. One of the greatest lessons of all for our renewable future is to learn how to stop wasting energy. Energy conservation and efficiency is our most precious and forgotten renewable energy.

Can we leave the Fire Age behind? Will electricity supersede fire in time? The answer may determine our future on Earth.

WALT PATTERSON,
Electricity Vs Fire (2015)

09 NEGATIVE ENERGY IS POSITIVE

'Every act of energy conservation ... is more than just common sense: I tell you it is an act of patriotism.'

PRESIDENT JIMMY CARTER (1979)

We could power *two additional* planet Earths on the energy we waste each and every day: one extra planet on the waste from power stations, and the other on the waste from the poor design of our transport systems, buildings and appliances. This three-planets story is a nice memorable fact but, if anything, it is an understatement of the true scale of how wasteful our energy systems currently are. We've seen in previous chapters that we can measure both the quantity and the quality (the potential exergy to perform useful work) of an energy source. The smartest design engineers measure exergy. So, if we put on a pair of 'exergy goggles', the waste in the global energy system is closer to 90 per cent.

Think about the tens of trillions of dollars we spend, the wars we fight, the billions of tonnes of carbon dioxide we emit, and the myriad environmental impacts associated with our global energy supply systems of coal mines, oil and gas fields, pipelines, power stations and electricity grids. Now think about the gadgets, buildings, industries, cities and transport systems that we have designed to guzzle these energy flows on demand. We've built an energy supply system and energy demand system that somehow manages to waste 90 per cent of the true thermodynamic potential of the energy flowing through it. Through exergy goggles, we don't look that smart as a species.

If there were no negative side effects of gathering and using energy, or no 'externalities', as economists say, the waste wouldn't matter so much. As long as we lived in a world of magical, abundant, cheap, benign energy, who would care? Some scientists continue to work for that day. But, until it comes, there are serious externalities associated with all energy supply sources, including renewables, and so the waste really does matter.

Low carbon, low energy, high renewables

Most of us will be familiar with the idea that if we are to avoid dangerous levels of global heating, we need to transition to low- or zero-carbon economies by the middle of this century. What we may not know is that the only real way to achieve zero-carbon economies is to radically increase our dependence on renewable energy supplies while simultaneously reducing the amount of energy we waste.

We also need to stop wasting fossil fuels. At our current rate of guzzling them, we have about ten years of their services left, if we are to stay below the Paris Agreement's 2 °C (3.6 °F) threshold. However, we need to stop wasting *all* forms of energy, including clean renewable sources of heat and electricity. The sooner we shrink our total overall demand for energy, the sooner renewables will be able to provide 100 per cent of the energy we need to power our zero-carbon economies. Scientists might yet invent safe sources of electricity and heat that are too cheap to meter, but until that wondrous day we'll need to manage global energy demand and consumption so that our renewable energy systems can cope.

Energy saving is a kind of fuel: a source of energy. Negative energy can be positive. The decade after the 1973–74 oil crisis, when the Organization of Petroleum Exporting Countries (OPEC) quadrupled oil prices almost overnight, saw a global boom in interest in the belief

> **We are in a race to get costs down as fast as possible, so as to avoid having to face much higher costs if we fail to expand renewables and efficiency fast enough.**
>
> DAVE ELLIOTT, Professor of Technology Policy, Open University (2020)

of energy conservation and energy efficiency. Reducing oil imports became a top priority in many oil-importing nations of the world. In the decades that followed, prices eased, supply increased and fears about energy security dwindled. What we didn't know was that while we may not have been running out of fossil fuels anytime soon, we were running out of atmosphere in which to dump the combusted carbon dioxide. So, how can we radically reduce our energy consumption?

Imaginary energy

One of the easiest ways to reduce our energy consumption is to get our international businesses, organizations and governments to stop cooking the energy books. For nearly sixty years, fossil-fuel companies, national utilities and national and international statistical agencies have been indulging in the creation and sharing of statistics that are, in part, based on completely imaginary sources and forms of energy.

The concept of 'primary energy', as it is known, measures 'raw' natural energy inputs into the economy before humans refine them. It has become one of a handful of key performance indicators of national and global trends in energy dependence, energy efficiency and energy futures analysis. In the strange world of primary energy accounting, one unit of electricity from a nuclear power station is commonly written down in the balance sheet as three units of energy at the primary level. The nuclear electricity is imagined to have come from a coal-fired power station. Some statistical sources apply similar imaginary factors for renewable electricity, others not. As the world shifts away from fossil-fuel-generated electricity towards modern renewables, the imaginary world of made-up primary energy numbers is more than a statistical or philosophical difference in the way we define and account for energy. It blinds electorates, policy analysts and decision-makers to the real picture of energy consumption and trends. It obfuscates reality and suits the fossil-fuel industry and anyone else interested in arguing that a sustainable energy future is too hard to design, too radical, too risky, too expensive.

Many future energy scenarios still use primary energy as their main measure despite the fact that the fictional elements of the raw resources in the underlying energy balances start to grow significantly in any low-carbon future scenario. Shell, for example, is well known for its excellent and methodical work on energy futures and regularly publishes a set of future energy scenarios. It has a lower energy, high efficiency, higher renewables scenario, known as Sky, to hold the increase in the global average temperature to well below 2 °C (3.6 °F).

If we examine the primary energy demand over the course of the century under Shell's Sky scenario using the different accounting systems, we find that energy consumption in 2100 is either 850, 1,050 or 1,400 EJ per year. All three numbers describe *exactly the same future* – they just use different accounting methods. But what's the true task we face? Are we looking for 850 or 1,400 EJ of renewable energy by 2100? Thankfully, possibly the most credible authority in the business, the Intergovernmental Panel on Climate Change (IPCC), uses the most physically accurate method, suggesting we need to aim for the much less challenging 850 EJ estimate. On our journey to a zero-carbon, 100 per cent renewable energy future, we must make sure to stick to the physics of the direct energy content that is being generated, transmitted and used.

Energy return on energy invested

Renewable energy is sometimes accused of doing more harm than good. Some sceptics suggest that certain renewable energy technologies use up so much energy to set up and operate compared with what they collect over their lifecycle that they might not be worthwhile. The second law of thermodynamics tells us that it takes work and effort to gather, concentrate and transform energy from one form into another, so it follows that we always have to expend energy to gather energy. And not all sources of energy are as easy to access as others. As noted earlier, the ratio of the energy gathered divided by the energy invested to do so is called the energy return on energy invested (EROEI). Ideally, we want the EROEI ratio number to be large. We don't want to spend a lot of money, time and actual physical energy gathering energy.

Largely depleted sources of 'easy-to-get-at' oil and gas had very high EROEIs, above 100. As oil and gas exploration has had to go further offshore and deeper, the EROEIs have been rapidly declining. A lot of energy, for example, is used in extracting and cleaning up tar sands and oil shale, compared with lifting Saudi Arabian sweet crude out of the ground. There's a lot of coal still left, but it is often very hard to get at and sometimes very poor quality. Coal, oil and gas can have EROEIs well below 10. Concentrated solar, tide, wave and nuclear power are all likely to be below 10. Large hydro schemes can be as much as 200. Recent academic papers suggest that the current EROEIs of wind and solar photovoltaics are generally high (≥ 10) and increasing, while some biomass energy conversion processes such as heating technologies and some ethanol and biodiesel fuels can have very low EROEIs.

Making sure we get enough return on the energy we invest is certainly a wise thing. The returns aren't purely about energy, but include reduced climate risks, energy diversity, resilience as well as economic and social benefits. Globally, the world remains addicted to fossil fuels. If we were to stop using them completely and immediately, as is sometimes argued for, it would impact on our ability to build the renewable energy technology system that we need for the future. We must use the remaining budget we have left of fossil fuels (to stay below 2 °C or 3.6 °F) to build the smart, more efficient buildings, factories and transport systems that we need for a low-energy, high-renewable future. Once we have reduced our energy wastage and increased our production of renewables, we will eventually reach the point where the future maintenance and replacement of the energy system can be entirely dependent on renewable energy. The sooner we stop wasting energy and using the remaining fossil-fuel energy productively, the sooner the day will come when the world runs more or less on 100 per cent renewables.

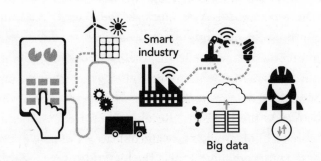

Energy intensities

Hunter-gatherers used about 6 GJ/capita per year of food and fuel in their diet. Averaged over the total human population, each of us today uses 77 GJ per year, although there are large differences between countries.

The average Chinese and UK citizen uses about 100 GJ. The US citizen uses nearly 300 GJ. Qataris use 650 GJ, and in the number-one slot is Iceland, where the citizens on average use a whopping 720 GJ each per year. Every country has a particular relationship with energy consumption and this depends on a large number of factors, including its climate, its history of development and its local available energy resources. Qatar, a major oil and gas exporter, gives away cheap energy to its citizens. Iceland has prodigious supplies of renewable geothermal energy, which get exaggerated in the primary energy accounting fiasco.

The good news is that it seems to be a universal truth that post-industrial economies experience a reliable, continuous year-on-year improvement in their energy productivity. In other words, their economies grow more quickly than the total amount of energy that they consume. Somehow, magically, things just keep getting better and better. As the Industrial Revolution evolved, we became more technologically savvy and able to get more work out of the same amount of fuel. The high tide of energy waste for the US was in the 1920s, China in the 1970s, Russia in the 1990s and India in the early 2000s. When economists compare the overall amount of energy that an economy uses in a year and divide this, for example, by the amount of economic output measured as GDP, they call this the 'energy intensity' of an economy.

> We must make every effort to save energy and focus on using various sources to meet our energy needs.
>
> ANGELA MERKEL (2006)

Since 1990, global GDP has more than doubled whereas total primary energy supply grew by 60 per cent. This is because technologies get better over time. LED lighting is perhaps the most spectacular recent example. LEDs produce the same lumens at much lower power consumption than their predecessors, compact fluorescents and incandescents. Some of these energy-efficiency improvements are in part because governments in many parts of the world have introduced energy-efficiency policies and regulations. When energy modellers build future scenarios, they build in the expectation that things will get better. They call it the 'autonomous energy efficiency improvement' factor.

If the global energy intensity was not in decline, we would be using two or three times the amount of energy we currently do. But we can, and must, do much better and increase energy efficiency much more quickly. Just a third of the world's final energy consumption is covered by energy-efficiency policies or regulations. Two-thirds is not.

The whole world doesn't simultaneously buy a new car or washing machine. We all don't knock down our old buildings overnight and replace them with energy-efficient modern ones. When new gadgets, cars, buildings, machines and factories are added to the economy, they are slowly replacing an existing stock of those things. The effect of any new breakthrough efficiency technology is watered down by the existing stock of gadgets. In recent years, for example, US energy intensity has declined at an average rate of 3 per cent per year. This was driven in part by a 5 per cent increase per year in efficiency of new devices and technologies. To reach a lower-energy, lower-carbon, higher-renewable energy future, we need some more radical technologies and, more importantly perhaps, some more radical ideas. The good news is that we can do more. A lot more.

Negawatts

The term 'negawatt' was coined by Amory Lovins, the co-founder and chief scientist of the Rocky Mountain Institute. He is one of the world's best-known and respected energy analysts and arguably has done more than anyone to reduce global energy demand and oil dependence.

A negawatt is a watt of energy that you do not use either because you have switched something off or you have replaced that something with a more efficient something else that uses less power. Negawatts

> Let's face it, the cheapest energy is the energy you don't use in the first place.
>
> SHERYL CROW (2007)

are real. If enough homes, offices and factories switch off, turn down or replace their lighting, heating and cooling gadgets, the macro effect is a significant reduction in electricity demand on the grid and therefore less need to burn coal or gas. If enough of us turn off the lights, the power grid notices and responds. A spectacular example of this was during the Covid-19 pandemic in 2020, when many countries went into national lockdowns, shutting businesses and generating unprecedented amounts of negawatts, affecting electricity grids everywhere. In the UK on 28 June 2020, overall electricity demand fell to its lowest level this century, falling below 17 GW. It is normally about double this.

There is a quiet multibillion-dollar negawatt industry that has grown up in the management of commercial and public buildings. Sometimes referred to as energy service companies (ESCOs), their profits are due to some very easy double-digit returns on investment to be made by thinking

> Because climate change is more chronic and Covid is more acute, everything with Covid actually happens in a very, very short period of time, including the lessons learned.
>
> CHRISTIANA FIGUERES
> (2020)

carefully and in a structured way about how to reduce power demand.

Electricity users are becoming micro suppliers. The Internet of Things will allow many more appliances, vehicles, buildings and factories to be connected to the electricity grid in a demand-responsive mode. They will be turned on and off to help manage the load on the grid. As the share of variable renewable electricity on the grid increases, negawatts are a key tool in helping to manage the system.

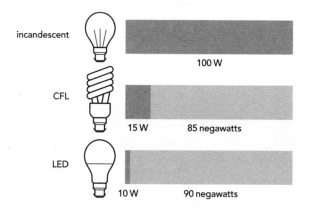

incandescent

100 W

CFL

15 W · 85 negawatts

LED

10 W · 90 negawatts

Energy efficiency

Efficiency is the ratio of an output to an input: what you get out divided by what you put in. The second law of thermodynamics applies, and the output is often less than the input. There is frequently a very long chain of events from getting a fossil fuel out of the ground and the delivery of what are called the end-use services: the thermal comfort, light or motive force that the energy user actually requires. Let's take the example of petrol used in a car. There are transmission and conversion losses at every step of the chain from the oil well to the car wheels:

1. Primary fuel in the ground (gushing oil and gas wells).
2. Secondary fuels (refined gasoline).
3. Customer delivery (getting it in the tank of the car).
4. End-use service (the passenger miles delivered).
5. Economic well-being (the economic and social benefit of the travel).

There are many ways to measure and report efficiency, depending on what is being divided by what. For example, three-quarters of the energy in a car's fuel tank is lost from the engine as heat. Less than 0.5 per cent of the energy in the fuel tank of a typical larger internal combustion engine car is used to move the vehicle. There are two major problems with cars: the nineteenth-century idea of using fire to create propulsion and the massive weight of the car compared with its occupant. Cars don't need to be like they are. It is the same for the myriad other energy technologies we use. It is technically possible to halve our global energy consumption by 2050 by shifting to renewables, but also by getting smarter on behavioural change and the design of our appliances, buildings and transport systems. And all without compromising the comforts we are used to.

Most importantly, the more we reduce our energy consumption, the faster our renewable energy systems will be able to cope. It will also help the current nearly 1 billion people who still don't have access to mains electricity, and the 2.5 billion who cook on open stoves with wood and cattle dung, as well as the additional 2 billion people who will have joined the human population by 2050, whose energy needs we must cater for.

When we dream of a sustainable society, it is usually electric. So we need to stop taking energy for granted, and we've got to stop wasting electricity in particular. None of this is going to happen naturally – it requires some political determination, and the key to that is imagination.

10 IMAGINATION IS RENEWABLE ENERGY

'What if it's all a hoax and we've created a better world for nothing?'

NAOMI KLEIN (2014)

The most powerful source of renewable energy is our collective human imagination. Our twenty-first-century globally sustainable energy economy will be driven by good design, complex systems management and a cornucopia of modern technologies and behaviours.

Let's recap the story so far. Not all forms of energy are equally useful. Electricity and mechanical energy are higher-grade forms of energy and have more potential than lower-grade forms, such as heat. Globally, the amount of renewable energy we can harvest is many more times greater than our present and utterly wasteful rate of energy consumption. This is just as well because our fossil-fuel party is all but over. The carbon clock is ticking. We could, if we so wished, live on sunshine alone. There's more than enough sunlight to go around. The solar PV and CSP industries are on course to be twenty times bigger than they are today. Wind power can meet around a third of global electricity supply – the industry is on course to be at least ten times bigger than it is today by 2050.

Most of the renewable energy we produce today is from modern bioenergy and hydropower. Proven and reliable, they are descendants of ancient technologies and both will continue to power a good chunk of our energy needs into the future. Deep in the ground and the across the oceans there are almost unfathomable amounts of heat ready to be tapped and converted into electricity. Harnessing energy from extreme environments is nothing new and entirely possible. Ask a coal miner, or an offshore oil and gas rigger.

In the next two decades, electricity use will increase by nearly 50 per cent. To stay well below a 2 °C (3.6 °F) increase on pre-industrial temperatures, renewable electricity output is expected to increase from 26 to 68 per cent, driven mainly by wind and solar power, with more modest yet still impressive increases in hydro and other renewable technologies.

Depending on what sort of energy goggles you put on, our current energy system wastes between 66 and 90 per cent of the energy we fight for, and fight over. The future is electric and renewable, but to be low carbon it must also be low energy. We need to be imaginative in our politics, engineering and economics to make this future happen.

Global imagination

The mainstreaming of renewables hasn't all been plain sailing. Anti-renewable NIMBYs have groaned about covering the land and seascape with ugly, expensive, unreliable and inefficient wind turbines and solar farms. More informed sceptics tend to do some lifecycle analysis to reveal very low energy returns on energy-invested ratios for renewables in an attempt to sow the seeds of doubt about our ability to rely on variable renewable energy on the grid.

All forms of energy capture, ancient and modern, have economic, environmental and energy costs associated with

> **It is a feature of renewable energy sources that their costs and benefits are *congruent*: they accrue to more or less the same populations; while for fossil fuels the costs and benefits fall on *different* populations. Renewables are simpler, fairer.**
>
> PETER HARPER, Centre for Alternative Technology (2020)

them. There are, alas, currently no magic energy bullets. And not all forms of renewable energy are equal or suitable for all places and applications. Specific technologies have their advantages and drawbacks, depending on the context. The sustainable energy future for many countries will be a carefully selected patchwork of technologies appropriate to their national circumstances. One thing, however, is common to all countries and all visions of our energy future: there is no way that we can all live like today's average American, Brit or Middle Eastern. Happily, there will be no need to, either.

In the last few years, a number of credible, detailed quantitative scenarios that meet the goals of the 2015 Paris Agreement (to stay well below the 2 °C or 3.6 °F increase in global heating compared to the pre-industrial era) have been published. They have been modelled by a broad range of organizations, including BP, the Intergovernmental Panel on Climate Change, the International Energy Agency, the International Renewable Energy Agency, Exxon, Greenpeace, Shell, Statoil, the US Energy Information Administration and the World Energy Council. While all these scenarios by no means paint exactly the same picture of what a sustainable energy future looks like, they do share a number of features that give us a reasonable insight into the sorts of changes we are likely to see in the coming decades:

- All scenarios include rapid shifts away from fossil fuels towards modern renewables (wind and solar in particular) to generate both electricity and heat.
- In many scenarios, any remaining use of fossil fuels is assumed to be associated with some form of commercial-scale carbon capture and storage technology (using fossil fuels, but cleaning up their mess).
- Most scenarios assume a significant decoupling of the growth in energy demand from the growth in GDP. In other words, they assume low-energy futures with a doubling or even trebling of energy efficiency.
- Many scenarios show large-scale land-use changes around agriculture for food and fuel as well as forestry to sequester significant quantities of carbon.

Generally, they all depict a decline in global fossil emissions to net zero by around 2050 to 2075. Many of them assume significant carbon dioxide removal (negative emissions technologies) in the last half of the century. The old fossil-fuel companies, perhaps naturally, tend to see a slower decline in coal, oil and gas markets aided by more faith in carbon capture and storage. Some scenarios emphasize a boom for nuclear, the hydrogen economy, carbon dioxide removal or all of the above.

In 2018, the IPCC published a comprehensive scientific review of how we might limit global heating to the more stringent 1.5 °C (2.7 °F) limit. The report examined socioeconomic uncertainties around our future energy system, but also took into account scientific uncertainties (of which there are many) about how the earth's climate system will respond. The report concluded that a sustainable energy future has the following generic features:

- A massive overall shift towards electrification. Electricity goes up from 20 per cent of final energy consumption in 2020 to between 34 and 71 per cent by 2050.
- A massive shift away from fossil fuels to renewable electricity generation. The share of electricity supplied by renewables increases from 26 to between 59 and 97 per cent.
- A significant change in land use for agriculture and forestry. There are reductions in the amount of land used for pasture and increases in the amount of agricultural land used for energy crops.

In general, the various 1.5 °C (2.7 °F) scenarios deliver a double whammy. They reduce the risks of global heating but also help alleviate poverty and improve public health through improved air quality, preventing millions from facing premature deaths.

Staying below 2 °C (3.6 °F) is not going to be easy and is certainly not anything resembling 'business as usual'.

However, the broad collection of sustainable energy scenarios shows that there are several believable pathways to meet our future energy needs and climate-change goals, and to achieve equitable access to safe energy for all. Renewable energy and energy efficiency are the stars of the show. They are mature technologies, ready to be deployed on a large scale. We may also need new, as yet unproven technologies, such as carbon capture and storage, BECCs or CDR (carbon dioxide removal) to help us get there.

The biggest elephant in the room is a lack of coordinated international political imagination and motivation to make this future happen at speed. Thankfully, some nations, cities and organizations are already leading the charge.

Political imagination

There are some clear, positive signs that international negotiators, national politicians, city mayors and leaders are willing to sign up to programmes and policies to switch over to energy efficiency and renewables. Around fifty-five countries have committed to some form of a 100 per cent renewable energy target by 2050 (some as early as 2030). In addition, around 280 cities, states or regions have committed to some form of a 100 per cent renewable energy target. The precise nature of the target is often not expressed – it can cover just electricity, or final energy consumption or primary. Iceland and Norway are already near a 100 per cent renewable electricity supply, with several more committed to the goal. Many nations are already more than 70 per cent sufficient in renewable electricity, including Brazil, New Zealand, Denmark, Ecuador, Venezuela, Austria and Columbia.

Within the EU, a number of countries have asked the European Commission to explore a 100 per cent renewable energy scenario to be added to their long-term planning horizon. The EU has a target of 32 per cent by 2030, Spain of 42 per cent by 2030 and Denmark leads the world with a 100 per cent renewable energy target covering its total energy supply for 2050. The majority of national regulations on renewables focus on the electricity sector. More than 140 countries have policies aimed at renewable electricity, about half of that covering transport fuels, with fewer than thirty countries with policies covering heating and cooling. A few government programmes have been established to support renewable hydrogen and are quickly gathering momentum. RE100 is a global initiative bringing together businesses committed to 100 per cent renewable electricity. Currently there are more than 260 members in 140 countries, with a combined electricity demand of 281 TWh/year – an electricity bill about the size of Korea's.

There are some positive signs, but collectively we remain on track for a 3 to 4 °C (5 to 7 °F) rise in global heating. We need more political imagination.

Engineering imagination

We currently spend very little on energy efficiency in our buildings, transport systems and industries. Surely, you might ask, if there are financially sensible ways of saving energy money and carbon, they would get done naturally? Well, unfortunately, no, they don't. This is because we have a rather strange way of doing the financial calculations around energy-efficiency investments.

The dominant mindset taught in most engineering schools is to examine improvements to a building or system in an incremental way, modelling one change at a time. Economists sometimes refer to this way of modelling as *ceteris paribus* – all other things being equal. When we financially model energy-efficiency improvements to a building, for example, we end up with a top ten list of the most cost-effective measures to take. We start with the measures that cost the least and save the most. In other words, we implement the things that get you the biggest bang for your buck, and then move to the next, and so on. As we go down the list, we reach a point where the energy savings get so expensive that we stop. This is the story of energy efficiency as a finite resource or a diminishing return.

There is a way of designing new things or retrofitting old things called integrative design, which allows us to access much greater energy savings that would normally and conventionally be uneconomic in a diminishing returns spreadsheet culture. According to its inventor,

Amory Lovins of the Rocky Mountain Institute (who coined the term negawatts), integrative design is the art of choosing, combining, sequencing and timing fewer and simpler technologies to save more energy at lower cost than deploying dis-integrated and randomly timed technologies.

Perhaps most simply put, integrative design is about buying energy efficiency rather than fuel. Where the traditional design rulebook comes up with an answer that says it isn't possible, integrative design emphasizes design over technology, rips up the rulebook and starts again with a redesign that finds ways of saving more energy overall with a greater financial return than the normal design. Amory call this 'expanding returns'.

Research and development

ARCHITECTURE

Project management

INTEGRATED DESIGN

ENGINEERING

DESIGN

A big part of the problem is that, just like negawatts, unused energy is normally invisible. For building energy efficiency, either by new build or retrofit, it is becoming more common to think about a 'whole building' approach. This is very sensible – do the right things in the right order. In many climates, this often means sorting out buildings' insulation as the first step and then moving on to think about heating, ventilation and air-conditioning (HVAC). By eliminating their heating systems and therefore their lifetime heating costs, passive houses, for example, cost about the same as a conventionally designed house. Saving capital costs by shrinking or eliminating HVAC equipment along with their lifetime operation

> **Here's where redesign begins in earnest, where we stop trying to be less bad and we start figuring out how to be good.**
>
> WILLIAM McDONOUGH AND MICHAEL BRAUNGART, *Cradle to Cradle: Remaking the Way We Make Things* (2002)

and maintenance costs pays for the additional investments needed to achieve radical efficiency gains.

The future is an energy-abundant society where we spend our capital on clever ways to harness the flows of modern renewable energy. We need to be more imaginative about how we stop buying fuel and electricity and start buying more energy efficiency in the form of better, smarter, cleaner renewable energy systems, buildings, transport systems and cities. The principles of integrative design can also be applied to the world of economics and finance. When we model the true full costs of alternative investments in our future energy system, the perceived additional costs of investing in renewable energy futures can radically disappear.

Economic imagination

Global Gross Domestic Product (GDP) in 2019 was around $US86 trillion. The world's annual fossil-fuel bill is around 10 per cent of its GDP. This is about $US10 trillion, an amount second only to global spending on healthcare. A significant part of the money we spend on healthcare is in turn spent on the health impacts of burning fossil fuels. A vicious circle. Imagine a global energy system running less on fuel and more on advanced, smart energy efficiency and cleaner renewable energy supplies, leading to cleaner air. A virtuous circle.

The top five wealthiest countries are the US, China, Japan, Germany and India, and they account for 55 per cent of the world's GDP. Interestingly, they are all serious producers and consumers of renewable energy. The long-term historical global GDP growth rate is around 3 per cent (of course, this is not evenly spread, with some countries, such as China, racing away and others, such as Japan, growing slowly) and most economic projections assume this will carry on to 2050, when GDP reaches about $US209 trillion. Annual global capital investments in energy supply capacity and incremental spending on more efficient equipment and goods are running at just under $US1.8 trillion. The biggest chunk by far (about $US0.72 trillion) is spent on oil and gas supply, a similar amount to our combined investments in power stations, nuclear power, renewable power and electricity networks. This is also about three times what we spend on energy efficiency. Global investment in energy efficiency remains relatively small at $US0.24 trillion. Capital investment on renewable power has reached $US0.3 trillion, which is about the same as we spend on the electricity networks and about three times more than we currently spend on investments in fossil fuel and nuclear power combined.

According to the International Renewable Energy Agency's 2050 Energy Transformation Scenario, $US3.2 trillion (about 2 per cent of global GDP) will need to be invested each year to limit global heating to 2 °C (3.6 °F). This is around $US0.5 trillion more than our current future plans. IRENA's analysis shows that increasing investments from $US95 to $110 trillion over the three decades, and rebalancing these towards energy efficiency and renewables, would bring higher overall economic growth worth at least $US50 trillion. In addition, it would also bring $US142 trillion worth of reduced climate- and air-quality-related health costs that would otherwise be absorbed by the financial and healthcare systems. The scenario also increases jobs in renewables to 42 million and overall jobs in the energy

> ❚ **Hundreds of people can talk for one who can think, but thousands can think for one who can see.** ❚
>
> JOHN RUSKIN, *Modern Painters Volume III* (1856)

sector to 100 million, up from a total of 60 million today. While some economists might argue with the precise numbers, IRENA's scenario is one of several credible financial assessments that all suggest the same broad returns on investment by accelerating the shift away from fossil fuels towards energy efficiency and renewable energy.

We need more imaginative economics. Well, to be precise, we need more political imagination around applying basic economics to create new political economies of thought. One idea in particular is the role of carbon pricing to send the right signals to the markets to encourage better investment decisions. Just 20 per cent of global greenhouse gas emissions were subject to carbon taxes in 2019. While some carbon markets are developing in Australia, the EU, Japan and Korea, much of the world and most of the greenhouse gases we emit are not priced correctly. Most 2 °C (3.6 °F) scenarios assume a significant increase in the carbon price in the next decades to around $US150 per tonne ($US136 per US ton) of carbon dioxide emitted in 2040/50. No taxes are popular, but imaginative approaches to carbon pricing can, and do, work.

Another area is funding basic research development and demonstration (RD&D); for example, of ocean, geothermal and advanced biomass technologies. Relatively small amounts of funding could be transformative for such technologies as ocean thermal energy conversion, enhanced geothermal systems or advanced biofuels. Globally, government spending on energy RD&D as a share of GDP is currently less than 0.05 per cent.

Our present global pattern of energy spending and capital investment resembles a tragically run investment bank. Despite all the scientific and economic testimony to the contrary, we continue to invest heavily in fossil fuels. All the evidence suggests that shifting fuel and capital expenditures away from fossil fuels and towards energy efficiency and renewable energy would not only pay for itself, but would bring considerably more long-term cumulative savings and profits, access to clean energy for all, more jobs, more income and better health. Reaching 100 per cent renewable energy needs to be our collective human mission. Imagine that.

> **Imagine fuel without fear. No climate change. No oils spills, dead coal miners, dirty air, devastated lands, lost wildlife. No energy poverty. No oil-fed wars, tyrannies, or terrorists. Nothing to run out. Nothing to cut off. Nothing to worry about. Just energy abundance, benign and affordable, for all, for ever.**

THE ROCKY MOUNTAIN INSTITUTE, *Reinventing Fire* (2011)

GLOSSARY

Active solar heating – the use of devices to harness solar power to produce hot water. Not to be confused with passive solar heating.

Bioenergy with carbon capture and storage (BECCS) – the CO_2 produced from the combustion of plants is captured and stored to reduce atmospheric concentrations of CO_2.

Carbon capture and storage (CCS) – the removal and prevention from entering the atmosphere of carbon or carbon dioxide from fossil or biomass fuels, either before or after combustion. Sometimes carbon capture, utilization and storage (CCUS).

Carbon dioxide removal (CDR) – the removal of carbon dioxide that is already in the atmosphere as opposed to pre- and post-combustion carbon capture and storage (e.g. by BECCs) or direct air capture.

Concentrated solar power (CSP) – the use of sunlight to generate high-temperature steam by means of an array of mirrors, which then drives a steam turbine to produce electricity. Most CSP plants can also store some thermal energy.

Direct Horizontal Irradiance (DHI) – measures sunlight that has been scattered by the atmosphere, sometimes called diffuse sky radiation.

Direct Normal Irradiance (DNI) – measures the vertical component of radiation coming directly from the sun.

Electrolysis – the chemical decomposition of a liquid or solution by the conduction of electricity.

Energy return on energy invested (EROEI) – the ratio of the energy gathered (by a system or technology) divided by the energy invested in order to gather the energy.

Entropy – a measure of the degree of disorder of a system.

Exergy – the amount of energy that is available to be used to perform useful work. The joule (J) is the standard SI unit of energy.

Final energy – the energy that arrives at the consumer's electricity meter or petrol pump after the losses in generation, refining and distribution. Sometimes called delivered energy.

First law of thermodynamics – the amount of energy within a defined boundary (or 'closed system', as physicists would say) remains the same. Other versions: energy is conserved; the total amount of energy in the universe is fixed; energy can be neither created nor destroyed. Energy can, however, be transformed into more useful or less useful forms. Essentially, the same as the law of conservation of energy.

Fuel cell – a device for producing an electric current in the reverse process to electrolysis – combining two gases (typically hydrogen and oxygen) to produce electricity.

Geothermal energy – heat drawn from the earth's internal heat.

Global heating – the increase in global average surface temperatures compared to 1850 (defined as pre-industrial) due to the anthropogenic greenhouse effect.

Global Horizontal Irradiance (GHI) – the total irradiance received by a square metre of surface horizontal on the ground. GHI = DNI + DHI × cos (z) (z is the solar zenith angle).

Global Tilted Irradiance (GTI) – the irradiance on a tilted surface, it can be calculated from GHI, DNI and DHI.

Gravitational potential energy – the potential energy that exists because of a height difference.

Greenhouse gas – any gas that is excited by (and therefore traps) infrared radiation and has a heating effect on the atmosphere.

Grey, blue and green hydrogen – hydrogen produced either from fossil fuels (grey hydrogen), with carbon capture and storage (blue hydrogen) or from electrolysis using renewable energy (green hydrogen).

Heat energy or thermal energy – a form of energy that is stored and transferred between systems with different temperatures.

Heat pump – a device that 'pumps' heat from a cooler region into a warmer one, providing either warming or cooling to a space. In an air-source heat pump (ASHP) the heat is taken from the air, while in a ground-source heat pump (GSHP) it comes from the soil.

Kilowatt-hour – a unit of energy. If a device operates with a power of two kilowatts for one hour, the total cumulative energy that has flowed in that hour is two kilowatt-hours (kWh).

Kinetic energy – the energy of a moving object.

Microgrids – groups of customers (loads) with distributed generation assets (wind farms, solar farms, gas power stations) that are managed as one distinct entity and are capable of being disconnected from the wider macro grid.

Negative emissions technology – bioenergy with carbon capture and storage, or BECCS, is the only renewable energy that offers the prospect of carbon dioxide removal or negative emissions.

Paris Agreement on climate change – the 2015 agreement at the twenty-first Conference of the Parties to the United Nations Framework Convention on Climate Change to hold the increase in the global average temperature to 'well below 2 °C (3.6 °F) above pre-industrial levels and pursuing efforts to limit the temperature increase to 1.5 °C (2.7 °F) above pre-industrial levels'.

Passive solar collection – the warming of buildings by solar energy via windows and the building fabric.

Photosynthesis – the process used by plants and other organisms to synthesize light energy into chemical energy from carbon dioxide and water. The chemical energy is stored in carbohydrate molecules, such as sugars.

Photovoltaic (PV) cell – light-sensitive, semi-conducting collectors that turn light into electricity.

Photovoltaic effect – the emission of photoelectrons into a vacuum from a material when electromagnetic radiation, such as light, hits it. Photoconductivity is the emission of electrons into a conductor from a material when electromagnetic radiation hits it.

Potential energy – energy that is stored, and that depends on the position within a force field (e.g. gravity of an electric field) of an object and not its motion.

Power – the rate at which energy flows, whose SI unit is the watt: 1 watt is 1 joule per second.

Primary energy – the total 'raw' energy content of an energy resource before that energy is transformed/processed.

Pumped storage – reservoirs used as water 'batteries'. Hydroelectric plants use their turbines to pump water uphill to the reservoir at night to enable generation at times of high demand in the morning.

Rated output – the power output of a particular generating device such as a PV panel or wind turbine under optimum conditions.

Second law of thermodynamics – the efficiency of any heat engine must be less than 100 per cent – or, in other words, the entropy of a closed system tends to increase with time.

Solar flux / solar irradiance / solar radiation / solar insolation – flux or irradiance describes the power or flow of solar energy per unit area; for example, an instantaneous reading of watts per m². When the flow is integrated over a period of time, such as a daily average over a year, kWh/m²/day, it is called solar irradiation or insolation.

Solar thermal – technologies that convert the sun's energy into heat energy.

Sustainable development – the idea that development is sustainable if it meets various economic, environmental and social sustainability requirements that balance the needs of future generations with those of the present.

Technical potential – the amount of renewable energy output obtainable by full implementation of demonstrated technologies or practices, before taking into account costs, or the need for policies to promote their adoption.

Thermal energy – the kinetic energy of an average atom of a gas at room temperature. More generally, heat energy.

Watt (W) – the SI unit of power. One watt is an energy flow of 1 joule per second.

Work – the energy transferred to an object or system by a force displaced over a distance. The units of work are force × distance or the Newton × metre. Work is an orderly and higher-grade energy form. The standard scientific unit for measuring work is the joule. Power is the rate at which work is done.

Worldwatt – a unit used in this book for convenience to help calibrate the scale of different technical and market potentials of different renewable energy resources to the overall global primary energy consumption of the world economy. In 2019, one worldwatt was 599 exajoules per year or 19 terawatts.

FURTHER READING

For a detailed textbook review of the history, social, technical and economic aspects of all the main forms of renewable energy: *Renewable Energy*, by Stephen Peake (ed.), Oxford University Press (2017).

For an in-depth textbook review of sustainable energy systems: *Energy Systems and Sustainability: Power for a Sustainable Future*, by Robert Everett, James Warren and Stephen Peake (eds), Oxford University Press (2021).

For a brilliant analysis of what sustainable energy systems mean when the physics and the numbers are clear: *Sustainable Energy Without the Hot Air*, by David J. C. Mackay, UIT Cambridge (2009).

For an amazing account of the history of the emergence of the concept of energy: *Energy, the Subtle Concept*, by Jennifer Coopersmith, Oxford University Press (2010).

For the best account of energy transitions throughout civilization: *Energy and Civilization: A History*, by Vaclav Smil, MIT Press (2017).

For a classic early explanation of how energy efficiency and better planning can help achieve a low-energy future: *Soft Energy Paths: Towards a Durable Peace*, by Amory B. Lovins, Penguin (1977).

For a detailed examination of the role of economics in enabling the transition to low-carbon energy systems: *Planetary Economics: Energy, Climate Change and Three Domains of Sustainable Development*, by Michael Grubb, Jean-Charles Hourcade and Karsten Neuhoff (eds), Routledge (2013).

For a critical look at how renewable energy sources will cope in the service of net-zero global fossil emissions: *Renewable Energy: Can it Deliver?*, by Dave Elliott, Policy Press (2020).

For an exciting look at 100 per cent renewable energy systems: *100% Clean, Renewable Energy and Storage for Everything*, by Mark Z. Jacobson, Cambridge University Press (2021).

For a detailed insight into how the USA could power an economy in 2050 that needs no oil, coal, nuclear energy, one-third less natural gas and no new inventions: *Reinventing Fire: Bold Business Solutions for the New Energy Era*, by Amory B. Lovins and the Rocky Mountain Institute (2011).

For a comprehensive quantitative review of how energy efficiency and renewable energy technologies can eventually reduce greenhouse gases: *Drawdown: The Most Comprehensive Plan Ever Proposed to Reverse Global Warming*, by Paul Hawken (ed.), Penguin Books (2017).

INDEX